TECHNOLOGY, INNOVATION, and POLICY 2

Series of the Fraunhofer Institute
for Systems and Innovation Research (ISI)

Fraunhofer Institute for
Systems and Innovation Research (ISI), Karlsruhe, Germany:

Authors:

Rainer Bierhals
Dr. Stefan Kuhlmann (Project Manager)
Lionel Pilorget
Guido Reger (Project Manager)

Secretarial Work/Layout:

Helga Schädel
Christine Mahler-Johnstone

Statistics/EDP/Graphics:

Matthias Hildebrand
Sebastian Oppermann
Georgi Triantafyllou
Ralf Ziefle

Guido Reger · Stefan Kuhlmann

European Technology Policy in Germany

The Impact of European Community Policies upon Science and Technology in Germany

With 58 Figures and 56 Tables

Physica-Verlag
A Springer-Verlag Company

Guido Reger and Dr. Stefan Kuhlmann
Fraunhofer Institute for
Systems and Innovation Research
Breslauer Str. 48
D-76139 Karlsruhe, FRG

ISBN-13: 978-3-7908-0826-1 e-ISBN-13: 978-3-642-99778-5
DOI: 10.1007/978-3-642-99778-5

88/2202-543210 - Printed on acid-free paper

Foreword of the German Steering Committee to the IMPACT Study

Research and technological development are important basic preconditions for long-term competitive success of the economy in the light of international competition. Their support in the European Communities (EC) therefore occupies a more significant position than before in the Single European Act. And yet the share, consisting as it does of less than 4% of the EC budget, is still limited. It is widely doubted whether EC research support has attained its chief goal, that is, the improvement of European competitiveness on world markets. In Germany the EC research funding, due to the strong focus on information technology, has developed into an important factor in the efforts to increase the innovation potential of industry, in this field at least.

In the present study an attempt is made - as has already been done in other member states - to establish, substantiate, and assess this impact according to empirical methods. Subject of the study is the Second EC Framework Programme in the area of research and technological development (1987 - 1991). This represents, coming as it does on the eve of the Fourth Framework Programme (1994 - 1998), not only an accounting for the past, but should also create a basis for the necessary discussion on whether and in which points re-thinking, alterations, and possibly improvements are necessary and possible.

Interesting individual results confirm that the study was worthwhile: an improvement of the competitive position is more apparent in the Single Market than against Japan and the USA. Approximately half of the promotional funding is concentrated on 10 leaders in industry, higher education institutes and non-university R&D institutes respectively. The German participants concentrate on the programmes relevant to industry. Finally, the study brings to light the alarming result of the disproportionately low participation of German small and medium-sized enterprises.

The collaboration with the Fraunhofer Institute for Systems and Innovation Research carrying out the study and with the EC bureaux which commissioned the study was pleasant and unproblematical. Suggestions were discussed objectively and mainly taken up.

Thanks must be expressed to the Commission of the European Communities which commissioned this study and to the EC departments as helpful partners of the institute carrying out the study and the Steering Committee. The supplying of statistical data and other information on research support by the Directorate General for Science, Research and Development must be specially mentioned.

For the Steering Committee of the German Impact Study (Members):

RegDir Brand	Bundesministerium für Forschung und Technologie
Dr. Brebeck	Bundesministerium für Wirtschaft
Dr. Krull	Max-Planck-Gesellschaft (Generalverwaltung) (Wissenschaftsrat, Köln until February 1993)
Dr. Kunz	Wirtschaftsministerium Baden-Württemberg
Dr. Uhlhorn	Bundesministerium für Forschung und Technologie

TABLE OF CONTENTS

List of Abbreviations

AiF	Arbeitsgemeinschaft industrieller Forschungsvereinigungen Confederation of Industrial Research Associations
BIG	Large Enterprises
BLE	Blaue-Liste-Einrichtungen
BLK	Bund-Länder-Kommission
BMFT	Bundesministerium für Forschung und Technologie Federal Ministry for Research and Technology
BMU	Bundesministerium für Umwelt, Naturschutz und Reaktorsicherheit Federal Ministry for the Environment, Nature Conservation and Nuclear Safety
BMVg	Bundesministerium der Verteidigung Federal Ministry of Defence
BMWi	Bundesministerium für Wirtschaft Federal Ministry for Economic Affairs
CEC	Commission of the European Community
CREST	Scientific and Technical Research Committee
DFG	Deutsche Forschungsgemeinschaft e.V. German Research Association
DIHT	Deutscher Industrie- und Handelstag German Association for Industry and Commerce
DLR	Deutsche Forschungsanstalt für Luft- und Raumfahrt
EC	European Community
ECSC	European Community for Steel and Coal
FhG	Fraunhofer-Gesellschaft Fraunhofer Society
FRG	Federal Republic of Germany
GDR	German Democratic Republic
GFE	Großforschungseinrichtungen National Research Centres
GMD	Gesellschaft für Mathematik und Datenverarbeitung
HEI	Higher Education Institute
I&C	Information and Communication (Technologies)
IHK	Industrie- und Handelskammer Chamber of Industry and Commerce
IRDAC	Industrial Research and Development Advisory Committee
ISI	Fraunhofer-Institut für Systemtechnik and Innovationsforschung Fraunhofer Institute for Systems and Innovation Research
KFK	Kernforschungszentrum Karlsruhe

KoWi	Koordinierungsstelle EG der Wissenschaftsorganisationen Liaison Office of the German Scientific Research Organisations
MECU	Millions of ECU
MNE	Multinational Enterprise
MPG	Max-Planck-Gesellschaft Max Planck Society
OECD	Organisation for Economic Cooperation and Development
R&D	Research and Development
RDI	Non-university Research Institute
RTD	Research and Technological Development
S&T	Science and Technology
SME	Small and Medium-Sized Enterprise
SV	Stifterverband für die Deutsche Wissenschaft Donors' Association for the Advancement of Science
TU/TH	Technical University

1. Introduction

1.1 Purpose of the Study

When the Agreement for the Foundation of the European Economic Community (EC Treaty) was extended and changed by the Single European Act (SEA), ratified by all the Member States of the European Community (EC) in July 1987, research and technological development were placed for the first time on an equal footing with other political competences of the EC. In the newly added Articles 130 f - q of the EC Treaty the aims and policy measures are clearly formulated, thus giving the EC a real competence to act. The **main aim of European research and technological development policy** (RTD policy) is (according to Article 130f of the SEA)

- to strengthen the scientific and technical basis of European industry and the development of its international competitiveness,
- to support transnational cooperation between industry and science, and
- to integrate the area of research and technological development into the general concept for the realisation of the internal market of the European Community.

The main instruments of European RTD policy are the "**Community Framework Programmes**", running for several years, in which the aims, priorities and financial dimensions are defined for a period of four to five years. Up to now there have been three Framework Programmes, with a growing number of sub-programmes: the First Framework Programme (1984-1987) was followed in September 1987 by the Second Framework Programme (1987-1991)[1] and in August 1990 by the Third Framework Programme (1990-1994).[2] The Fourth Framework Programme will probably start at the end of 1994 and will run from 1994 until 1998. The overlapping of the programmes in time is deliberate, and is intended to ensure the continuity of the RTD activities of the Commission of the European Community (CEC).

The growing political and economic integration of western Europe, and the conferring of the competence to act in the area of RTD policy, have led to increasing importance of the EC as a policy agent in the German research landscape over the last 10 years. In the

[1] cf. Kommission der EG (1987): Amtsblatt der Europäischen Gemeinschaften L 302 vom 24.10.1987, Jahrgang 30.

[2] cf. Kommission der EG (1990a): Amtsblatt der Europäischen Gemeinschaften L 117 vom 08.05.1990, Jahrgang 33.

Federal Republic of Germany the federal government is the main actor in RTD policy. However, in the course of the historical development of German RTD policy **regional and supranational policy agents have** also **come to play an important role** (Meyer-Krahmer/Kuntze 1992, 97ff)). Today, as well as the federal government, the "Länder", local authorities, semi-public and private institutions (such as chambers of industry and commerce, transfer institutions, Confederation of Industrial Research Associations (AiF)) at a national or regional level, and the European Community at a supranational level, play an active role in the field of RTD policy in Germany.

In order to meet the growing challenges to RTD policy, and in consideration of the preparations for the Fourth Framework Programme, the **"IMPACT studies"** have been carried out at the request of the Directorate General XII (DG XII) of the Commission of the European Community (CEC) in order to assess the relative importance of European RTD policy for national policy in the Member States. Although there had previously been several studies and evaluations of EC RTD support programmes, these had analysed not the Framework Programme as a whole, but only a specific programme within the Framework Programme, or selected EC research projects. Moreover, the studies had aimed at an assessment of impacts at the European level and not at the level of individual Member States.

A systematic attempt at a horizontal, cross-programme empirical examination of the impacts of EC RTD support in the Federal Republic of Germany (and in the other Member States) has not previously been undertaken.[3] This is the reason for the IMPACT studies carried out in Germany and in almost all the other countries of the European Community, taking the Framework Programme II as their subject. The studies for some countries have already been completed.

ISI was commissioned by DG XII to carry out the study for the Federal Republic of Germany. The purpose of the IMPACT study is to assess the possible effects of EC RTD policy on the German research landscape; it is essentially concerned (as are the IMPACT studies for the other EC countries generally) with answering the following questions:

---

3 The most comprehensive attempt so far at a scientific consideration of EC RTD policy from the German point of view is the study by Starbatty/Vetterlein (1990), which deals with the subject from a regulatory viewpoint and examines the aims, instruments and programme implementation of EC RTD policy, and the division of tasks between policy actors; the (relative) importance, results and impacts from the point of view of German participants in EC programmes are not examined.

- To what extent, and in what support programmes, do German actors participate ? How can the **structure of German participants** be described as regards type of participant, position in the project consortium and regional distribution ?
- What are the **motives** and **aims** of enterprises, higher education institutes and non-university R&D institutes in participating in EC support programmes?

- What is the **significance** and the **relative importance** of EC RTD support programmes for actors in the German research scene?

- Do the research activities and strategies of the supported institutions coincide with the **aims of the EC support**? What **results** were achieved in the projects and what **impacts** have these had from the viewpoint of the participating institutions?

- What are the impacts of EC programmes on the **cooperation behaviour** of the participating institutions? Do they initiate, stabilise or lead to the subsequent expansion of transnational cooperation ?

- Does the participation in EC projects have a **value added**, which may not appear without participation or through projects in the national framework (**additionality** of EC support)?[4]

- What are the **reasons for non-participation** in EC research support programmes? What **difficulties** and **barriers** impede participation?

- What is the significance of EC RTD policy for German RTD policy, both at the **federal and Länder levels**? Do the area of support and the instruments used overlap or complement one another?

- What **interactions** are there between EC RTD policy and Federal/Länder policy? Is there an **exchange of knowledge and experience** between policy-makers?

- Does EC RTD support influence **other policy areas** such as environmental protection, energy and regional development?

The IMPACT studies carried out in Germany and in the other Member States of the European Community do **not** represent **evaluations** in the sense of detailed analyses and evaluation of the aims, instruments and impacts of EC RTD policy. The studies were not able to deal with questions concerning the underlying assumptions of the programmes,

---

4 For a description of the concept of **additionality** of EC projects see Tsipouri/Gonard/ Kuhlmann/Morandini (1992)

the suitability of the EC RTD programme for solving technical, economic or social problems, the strategic effectiveness of the EC programmes or the extent to which the desired target groups were reached. These aspects would have required the use of additional methods and an evaluation concept containing additional elements,[5] which would have exceeded both the means and the terms of reference of the IMPACT studies. Rather, the German study and the other IMPACT studies are oriented towards examining the relative importance and possible effects of EC RTD policy on the national research landscape and on national policy **from the viewpoint of the participants**.

In order to ensure comparability between the different national IMPACT studies, "harmonized specifications"[6] were developed for investigations in the Member States, establishing common, comparable formulations of issues but leaving room for country-specific aspects. The results of all the IMPACT studies are being combined in a **"synthesis report"** by the French institute AXION; for this reason we shall not explicitly consider in this study other IMPACT studies that have already been completed.

1.2 Subject of the Study

The study concentrates mainly on an assessment of the importance and impacts of the **"Second Framework Programme of Community Activities in the Field of Research and Technological Development (Framework Programme II)"**, which ran from 1987 - 1991 and had a total budget of 5,396 million ECU (EG-Kommission 1987). Other activities of the CEC in research and technology, such as the funding of cooperation in RTD beyond the European Community, and the related individual programmes outside Framework Programme II were not included in the empirical investigation. In this study it was only intended to analyse possible important interactions between these associated programmes and the Second Framework Programme; thus, **the following programme lines were explicitly excluded from the empirical survey:**

1. Education and training (ERASMUS, COMETT, LINGUA, TEMPUS),

2. Pilot projects and demonstration projects (e.g. THERMIE, Community environment actions),

---

5 For a detailed critical appraisal of the tasks, elements and methods of technology policy evaluation see Kuhlmann (1992).

6 National IMPACT studies of Community R&D Policy. Specifications for Future Studies.

3. Coal and steel research (technical and scientific research according to Article 55 of the European Community Steel and Coal Treaty (ECSC-Treaty),

4. Innovation and technology transfer (SPRINT, NETT),

5. Regional support of technological infrastructures (RTD programmes financed by the EC structural funds such as e.g. STRIDE or STAR),

6. Data and information transmission (e.g. CADDIA, TEDIS, INSIS),

7. Cooperation in RTD beyond the European Community (COST, EUREKA).

The Second Framework Programme was selected because approximately two-thirds of the supported projects have either reached their final stage or have been completed[7] and it is therefore possible for statements to be made on the project results (to some extent at least). Since the written questionnaires did not not always reach the appropriate contact partner, and since some addressees did not differentiate in their replies between the individual programmes of the Second and Third Framework programmes, or between the Framework Programme itself and the related programmes, a few of the answers (less than 2% of those received) refer to programmes outside the Second Framework Programme. Since these instances were concerned with technologically oriented R&D projects, they were included in the survey and assigned to the relevant programme lines.

The scientific players in RTD policy are the research-performing institutions themselves, which together make up the "**research landscape of the Federal Republic of Germany**" (Meyer-Krahmer/Kuntze 1992, 97f). In order to obtain differentiated statements about the importance of EC RTD support for the German research landscape, we distinguished the following three groups of "**R&D actors**" within this landscape, and a fourth group made up of "**policy actors**":

- **R&D actors** in the German research landscape:

  1. **Industrial enterprises** (here a difference was made between large companies **(BIGs)** with more than 500 employees and small and medium-sized enterprises **(SMEs)** with less than 500 employees),

---

7 According to the results of the written questionnaire conducted by ISI with German participants in EC projects within the Framework Programme II, 48 % of EC projects have been completed and 20 % are in their final phase; altogether, 571 establishments answered this question on the progress of the project.

2. **Higher education institutes (HEIs)** (higher education institutes/technical colleges),

3. **Non-university R&D institutes (RDIs)** (including, for instance, the Max Planck Society, national research centres, Confederation of Industrial Research Association institutes (AiF), institutes associated with higher education institutes, "Blue List" institutes, the Fraunhofer Society, and private R&D institutes; see chapter 6).

- **"Policy actors"** in the area of research and technology (including e.g. the Federal Ministry for Research and Technology (BMFT), the Federal Ministry for Economic Affairs (BMWi), the respective ministries of the 16 Federal Länder, associations, chambers of industry and commerce, transfer institutions, contact and liaison offices).

The three groups of R&D performers "Enterprises", "Higher Education Institutes" and "Non-university R&D institutes" taken together are subsequently referred to here as **"R&D institutions"**.

Regionally speaking, the study is confined to the **old Federal Länder**, as only two research establishments from the new Federal Länder took part in the Second Framework Programme. Nevertheless, in order to take into consideration the **special situation in the new Länder**, the written questionnaire addressed to research establishments which had not participated in EC RTD programmes was extended to include the new Länder.

1.3 Methods

The methodological instruments applied in order to carry out the task described above were **analysis of existing studies** on the subject, **data bank analysis** of German participants in the Second Framework Programme, **written questionnaires**, oral **interviews with experts** and the discusssion of results in a **Steering Committee**. As there was as yet no systematic, empirical analysis of the importance of EC RTD support for the German research landscape, written and oral questioning of the players in the German research scene constitutes a substantial part of the IMPACT study. The assessment of the importance and impacts of EC support is thus **reflected to a large extent in the subjective assessments of the players questioned.**

Analysis of the Literature

An analysis of the literature was made in order to investigate evaluations of, and studies on, RTD policy and individual support programmes of Germany and the EC, and was used as background material for evaluation of the empirical results. An examination of the existing studies on individual EC RTD programmes showed that although the studies have investigated the impacts of individual programmes at the level of the European Community, country-specific issues make up only a small part of the analyses or are not included at all. Increasingly, the individual programme evaluations commissioned by the EC have pointed out more emphasis should be placed on the aspect of interactions between the RTD support of the EC and the country concerned as regards programme generation and evaluation.

Data Bank Analysis of Structure of Participants

As yet no systematically coordinated data on the structure of German participants in the Second Framework Programme were publicly available. With the help of the AMPERE data bank of the CEC, ISI was able to establish the structure of the German participants with regard to the type of research establishment, size, position in the project consortium, regional distribution of participants and the distribution of projects within EC programmes of DG XII and DG XIII (cf. Chapter 3). Analysis is generally limited to the frequency of participation in EC projects (a total of 2,108 German participations in EC projects); it was only possible to use the financial contribution of the EC as an indicator in the area of DG XII. Unfortunately, for the programmes of DG XIII the financial contributions of the EC are only available for EC projects as a whole, but not for the shares of the German participants involved.

Written Questionnaire

The written questionnaire was addressed to enterprises, higher education institutes and non-university R&D institutes in Germany. These three types of addressees were subdivided into two groups which received questionnaires that were different in some respects but overlapped in others: the group of "**Participants**" in R&D projects within the Second Framework Programme and the group of "**Non-participants**", who had not yet taken part in an RTD support programme of the EC.

With the help of the EC's AMPERE data bank, we identified 2,108 participants in Germany who had taken part in projects under the Framework Programme II of the European Community (1987-1991). With the exception of two higher education institutes in

Saxony, participants were all from the old Federal Länder. As the project leaders could not be identified for all German participants, and the quality of the addresses was very uneven, it was not possible to contact all participants in EC projects.[8] From this basic population of **German EC participants**, 1,540 R&D institutions were addressed,[9] of which 621 answered (response rate of 40.3%); 35 of the questionnaires returned were unusable. Thus the analysis of the questionnaire is based on **586 replies**. Since most of the R&D institutions participate in several EC projects the questionnaires were not answered in all cases; the questionnaires were only completed for some selected EC projects as an example.

As a comparison group, a basic population of enterprises, HEIs and non-university R&D institutes which had **not** yet **taken part in EC projects** was compiled. In order to identify selection errors, a "participant" questionnaire was included with every questionnaire sent to non-participants - and indeed some R&D establishments in this group, mainly in the new Länder, did send replies stating that they had made successful applications and were involved in EC projects which had started at the beginning or in the middle of the year.

For the group of non-participant enterprises, a sample of ca. 1,200 R&D performing firms was extracted from the IKAS data bank of the Chambers of Industry and Commerce (IHK, Industrie- und Handelskammern) and the German Association for Industry and Commerce (DIHT, Deutscher Industrie- und Handelstag) consisting of 700,000 German enterprises entered in the Register of Companies (Handelsregister). With the help of the VADEMECUM data bank of the Donors' Association for the Advancement of Science in Germany (Stifterverband für die Deutsche Wissenschaft), we compiled a sample of about 800 departments and institutes at HEIs and RDIs that had not yet taken part in EC RTD support programmes.

Altogether we sent a total of 1,902 questionnaires to the **comparison group**, of which ca. **1,560** were addressed to **non-participants in the old Länder** and ca. **340** to **non-participants in the new Länder**. 578 R&D institutions replied, and of these replies 560

---

8 In order to obtain informative statements on EC projects and a high rate of response it is necessary to write specifically to participants in EC projects; knowing the name of the project leader is a prerequisite for this. Especially in large multinational enterprises (MNEs) and in HEIs, it is very difficult or impossible to identify the EC project without the precise name and address of the relevant project leader. Whereas the names of the main contractors could be found by going through the synopses of the relevant RTD programmes by hand, names of the partners and associated partners were not accessible. Unfortunately the completeness of the names and addresses of participants obtained from the EC was very variable.

9 1,805 questionnaires were despatched initially to EC participants; of these, 265 could not be delivered by the German postal service, or the addressees could not be identified within the establishment to which the questionnaires were sent.

questionnaires could be included in the analysis (30.4% response rate). Our intention in including R&D establishments from the new Länder in the comparison group was to take account of the possible future significance of EC RTD support for the new Federal Länder.

Interviews with Experts

In order to increase the reliability of the results, in addition to the written questionnaires over 60 interviews were conducted with actors from the German research and policy scene:

- **Enterprises:** twenty interviews were conducted with participating enterprises. The interviewees were leaders of EC projects, heads of R&D departments or R&D groups, administrative headquarters, managers or owners of SMEs; eight interviews were conducted with non-participating firms (especially SMEs) in the branches of mechanical engineering, electronic industry, software and the chemical industry.

- **Higher education institutes** (higher education institutes/technical colleges): Seven heads of university institutes and three EC representatives at higher education institutes/technical colleges were interviewed; when carrying out the survey we received quite a few written comments from university institutes whose applications had been turned down by the EC on the difficulties involved in submitting an application.

- **Non-university R&D institutes:** nine interviews were conducted with project leaders and experts from national research centres, the Fraunhofer Society, the Max Planck Society, the AiF (Confederation of Industrial Research Associations) and institutes associated with higher education institutes.

- **Policy agents:** a total of 17 experts were interviewed, from the Federal Ministry for Research and Technology (BMFT), the Federal Ministry for Economic Affairs (BMWi), the Ministry for Economic Affairs and the Ministry for Science and Research of Baden-Württemberg, the Ministry for Economic Affairs of North Rhine-Westphalia, Liaison Office of the German Scientific Research Organisations (KoWi, Koordinierungsstelle EG der Wissenschaftsorganisationen), the Value Relay Centre in Stuttgart (Steinbeis-Europazentrum), the EuroInfo Centre of the DIHT (German Association for Industry and Commerce) and the institutions of the specialist EC coordinators (nine EC programmes) in Germany.

Steering Committee

The IMPACT study has been supported by a Steering Committee which has provided expert advice during its conception and realisation. The Steering Committee was made up of experts from the Federal Ministry for Research and Technology (BMFT), the Federal Ministry for Economic Affairs (BMWi), the Federal Ministry for the Environment, Nature Conservation and Nuclear Safety (BMU), the Ministry for Economic Affairs of Baden-Württemberg, a representative of a large R&D institution and a representative from DG XII. During the course of the study a total of three meetings with the experts took place at which results were discussed and impulses received.

1.4 Structure of the Study

The present study begins with a presentation of the research landscape in Germany, a description of EC research support and an analysis of the interfaces between RTD support by the German Federal Republic and by the EC (**Chapter 2**). In **Chapter 3** the structure of German project participants in the Second Framework Programme is analysed. In the subsequent chapters the motives, aims and results of EC RTD support and its impacts on the research actors "enterprises" (**Chapter 4**), "higher education institutes" (**Chapter 5**) and "non-university R&D institutes" (**Chapter 6**) are elaborated on the basis of the written questionnaires and interviews. **Chapter 7** deals with the specific situation in the new Federal Länder. In **Chapter 8** the most important results for the various actors involved are summarily evaluated and are related at a higher level to German and European RTD policy. **Chapter 9** closes the study with a look ahead at the Fourth Framework Programme and the possible future role of EC RTD support.

2. The RTD Policy of Germany and the European Community

Before analysing the importance of EC RTD support for Germany, it is appropriate to describe the German research and innovation system and present its actors.

Section 2.1 gives an outline of the German research and innovation infrastructure and of the policy directed towards it. Section 2.2 gives a short account of EC RTD policy, before going on in Section 2.3 to deal with the interfaces between German and EC policy. Here it emerges that the relative importance of EC RTD policy is (still) comparatively limited, that until recently the German actors were relatively restrained in their participation in the shaping of EC RTD policy, and that it is only in the last few years that EC support is of growing importance for an increasing number of R&D institutions.

2.1 The German Research Landscape

In international comparison, the German research and innovation system is considered to be **well developed**: in 1991, 425,000 persons performed research and development (full-time equivalent).[10] Altogether, expenditure on research in Germany in 1992 amounted to 80.7 thousand million DM. As a percentage of the gross domestic product (2.7%) this still puts Germany among the world leaders, but in the last few years it has fallen somewhat behind Japan (3.8%) and the USA (2.8%) (cf. BMFT 1993).

The German R&D system is regarded as **highly differentiated**: Figure 2.1.1 shows the most important groups of institutions in the German research landscape. Along the horizontal axis, the R&D institutions are grouped according to their main source of financing (between the poles of public/private): by far the largest group is constituted by the mainly privately financed R&D activities of the industry. There is also an intermediate group of institutions which cooperate closely with industry, but also have public sources of financing; these include the institutes of the Fraunhofer Society (FhG) and the laboratories of the Confederation of Industrial Research Associations (FuE-Einrichtungen der Arbeitsgemeinschaft industrieller Forschungseinrichtungen - AiF). The other institutes represented in the figure (Max Planck institutes, national research centres, federal and

---

10 These statistics refer to the Federal Republic of Germany in its previous form. Estimates indicate that in 1992 nearly 475,000 persons were employed in R&D in the whole of reunified Germany, including the New Länder. However, this figure is not considered completely reliable; reliable official statistics will only be available from 1993 (cf. BMFT 1993). Further detailed information on the research and innovation system in the New Länder is given in Chapter 7 of this report.

Figure 2.1.1

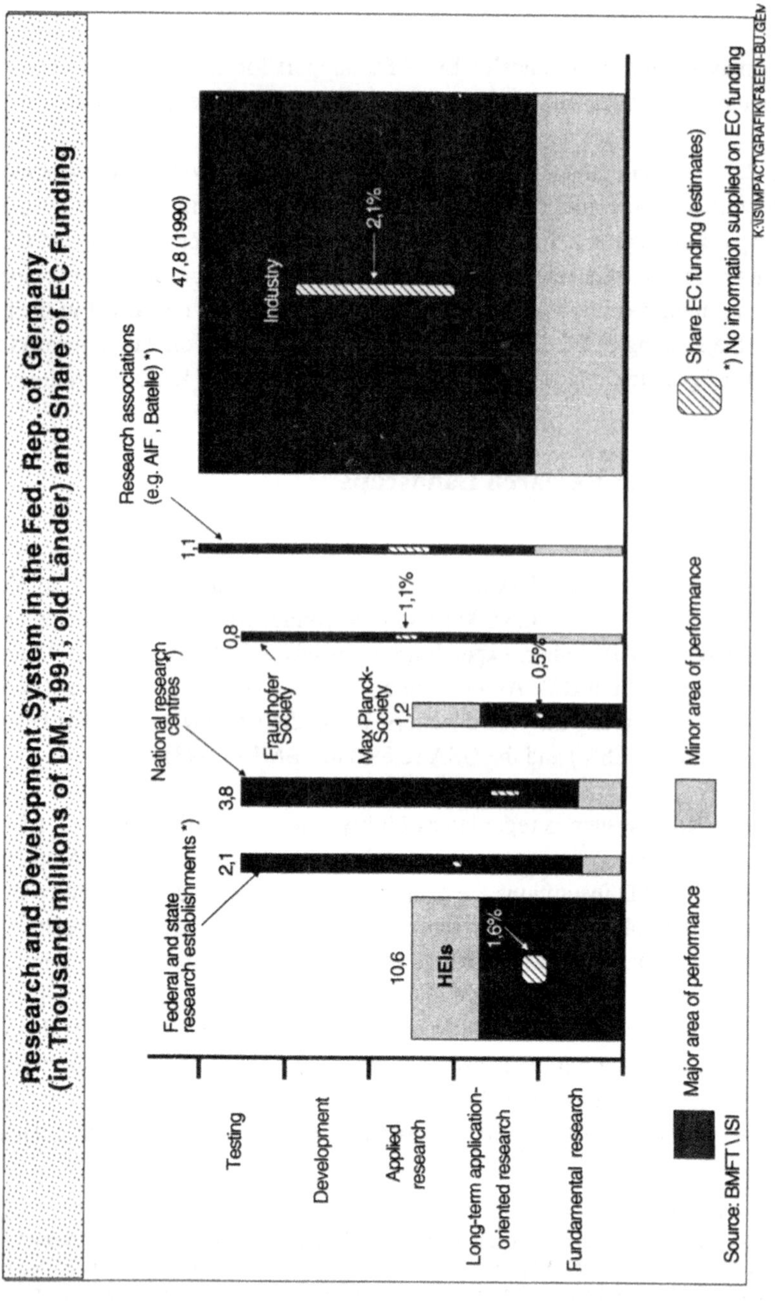
Research and Development System in the Fed. Rep. of Germany
(in Thousand millions of DM, 1991, old Länder) and Share of EC Funding
Testing
Development
Applied research
Long-term application-oriented research
Fundamental research
Federal and state research establishments *)
2,1
3,8
National research centres *)
10,6
HEIs
1,6%
Fraunhofer Society
0,8
Max Planck-Society
1,2
0,5%
1,1%
1,1
Research associations (e.g. AIF, Batelle) *)
47,8 (1990)
Industry
2,1%
Major area of performance
Minor area of performance
Share EC funding (estimates)
*) No information supplied on EC funding
Source: BMFT \ ISI

state research establishments (Bundes- und Landesforschungsanstalten), higher education institutes) are predominantly publicly financed. The vertical axis of Figure 2.1.1 groups the institutes according to types of R&D: basic research, application-oriented research with a long time horizon, applied research, development, testing. It emerges clearly that the national research centres and the Max Planck institutes are active mainly in the fields of basic research and long term applications-oriented research, whereas the work of industry lies primarily in applied research, development and testing. Only relatively few large multinational enterprises (MNEs), especially in the chemical and electrotechnical industries, conduct long-term, applications-oriented research in some fields, but with a shorter time horizon. The sizes of the fields in Figure 2.1.1 represent the approximate shares of the groups of institutions in total expenditure on R&D in Germany. It is clear that:

- **Industry** accounts for the **major share of R&D in Germany**: in 1990, almost 70% of expenditure on research and development was performed by industry (nearly 62% of R&D expenditure was financed by industry). In Germany, applied research and experimental development are almost exclusively the province of industry.

- The **higher education institutes (HEIs)** have the **second highest research budget** (the data in Figure 2.1.1 do not include expenditure on university teaching). The HEIs concentrate on basic research and long-term application-oriented research; in the course of the last 10 years, however, there has been a marked increase in the share of industrial research and development in the research budget of several HEIs, especially technical higher education institutes.

- The **federal and state research establishments** (Bundes- und Landesforschungsanstalten) carry out research, so-called "Ressort-Forschung" which is more or less directly associated with the tasks of the public offices that finance them, usually federal or "Länder" ministries (e.g. the development of norms and standards, testing procedures, certification etc.).

- The **national research centres** (Großforschungseinrichtungen/GFE; of which there are 16 at present) perform R&D which cannot be carried out either by the higher education institutes, the Max Planck Society, the industrially oriented research institutes or by industry itself. This includes particularly activities with a **long-term orientation** which appear to have a **high degree of risk**, require relatively **large research teams** and incur high costs; the national research centres typically perform research which requires **very large technological plants**. In recent years the national research centres have diversified their fields of activity considerably and are now active in a wide range of topics, especially in high energy physics, space technology, medicine, biotechnology, applied mathematics, software technology and environmental technology.

- The institutes of the **Max Planck Society (MPG)** mainly perform **basic research** in selected fields of natural sciences and the arts. They concentrate on new research topics of potential future importance which either have not yet found an established place at higher education institutes, or - because of their interdisciplinary character or the resources they require - cannot find a place there.

- The institutes of the **Fraunhofer Society (FhG)** have the task of promoting the practical application of scientific knowledge through **long-term application-oriented and applied research**. The Fraunhofer Society performs mainly contract research which is financed partly by industry and partly by the public sector. The FhG regards itself as the **"intermediary" organisation between science and industry in Germany**. No other German R&D institution has grown so rapidly over the last 10 years.

- The laboratories of the **Confederation of Industrial Research Associations (AiF)** perform tasks in **applied research and experimental development** for the sector-specific needs of industrial enterprises. Their work, which is financed partly from public sources and partly by industry, particularly benefits SMEs which are grouped into sector-specific research organisations.

One of the main reasons for the extent of institutional differentiation in the German research landscape is the fact that there is no strong, central policy body responsible for science, research and technology: within the federal system of the Federal Republic of Germany, it is essentially the Federal Länder that are responsible for academic and research aspects; although in the course of time the Federal Government has, in agreement with the Länder, taken on several areas of responsibility, there is always a certain degree of competition between the Federal Government and the Länder on the one hand, and between the various Länder on the other. The Länder are responsible for the higher education institutes; they maintain (to a varying extent) non-university research institutions and - also to a varying extent - have their own technology policy programmes. This does lead to some redundancies in the capacities of the research and innovation system, but it also guarantees a **decentral, "autonomous" structure of research capacities**, even outside the large cities.

Against this background, it is not surprising that German RTD policy relating to industry over the last decade is described as "diffusion oriented" rather than "mission oriented" (cf. Ergas 1987, 192). A strategic industrial policy with definite missions, such as is pursued in France, can hardly be implemented in a decentrally structured research system. Another reason for the stronger **"diffusion orientation" of German RTD policy** is the basic attitude prevalent in public policy that the state does not wish to be regarded as a "director" of industry, but only as the guarantor of an effective academic training system (higher education institutes), a high level of basic research and a sound technology base for industry (industrially oriented RTD programmes). However this basic position, as critics continually point out (cf. Starbatty/Vetterlein 1990), only works out to a limited extent in practice.

Since the 1950s, German RTD policy - as in other western European industrialised countries - has **continually extended the scope of its activities** (Meyer-Krahmer/Kuntze 1992). Bräunling and Maas (1988) have illustrated this as a "shell model" (cf. Figure 2.1.2):

- In this model, the core area of RTD support in the **1950s** comprises the **(basic) university research** financed by the Länder and the special area research carried out in federal and state research establishments (so called "Ressortforschung"). The first "layer" is characterised by **big technology programmes from the mid-50s onwards**, showing a marked orientation towards goals similar to those of the USA, mainly in the fields of nuclear technology, aerospace and data processing, and later microelectronics.

- A second research support "layer" was developed - according to Bräunling/Maas (1988) - from the **beginning of the 1970s**, in order to create and support the conditions necessary for the export of technology-intensive goods. Public funding flowed into research projects of **industry** and **institutes for applied research** to promote **cross-sectoral technologies** (e.g. materials), **key technologies** (e.g. microelectronics) and **technological systems** (e.g. transport systems).

- In the 1970s, the **reform policies** of the government of the Social Democrats triggered the formation of a third research support "layer", complementing the goals of the first and second layers by research activities in the areas of environment, public health and the employment market whose **aims** related primarily to **social policy**.

- From the **end of the 1970s**, a fourth "layer" came into existence with the instruments of **innovation policy**: their aim is the diffusion of innovative or improved technologies, also among SMEs and in less developed geographical regions. This includes the support of activities in R&D as well as the building up and strengthening of an infrastructure for the support of technology transfer from the science system into industry.

Figure 2.1.2: A Shell Model of Public R&D Support

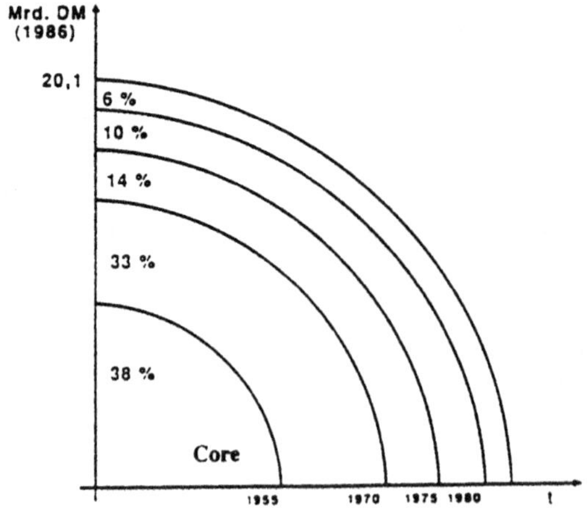

Core:	Basic Research
	Special Area Research
1. Layer:	National Research
2. Layer:	Key Technologies
3. Layer:	Employment and Environment
4. Layer:	Innovation

Source: Bräunling, Maas 1988

The shell model clearly shows that over the last 40 years these focal points of policy have not succeeded one another, but have **accumulated!** The result has been the formation of a **differentiated range of instruments** - sometimes described as "lacking transparency" (Kuhlmann 1992) - in public RTD policy (cf. Figure 2.1.3), extending from institutional support measures, over financial research support programmes and the formation of institutions for technology transfer, to regulative policy measures (e.g. competition policies) (cf. Meyer-Krahmer/Kuntze 1992).

Figure 2.1.3: Instruments of State Technology Policy

In a narrow sense	In a broader sense
1. Institutional Promotion - National research centres - Fraunhofer Society - Max Planck Society - Higher education institutes - Other institutions	**4. Public Demand** Use of public institutions to increase market pull for "desirable" technical developments, e.g. environment-friendly consumer goods (recycling paper, "eco" car)
2. Financial Incentives - Indirect promotion - Indirect-specific promotion - R&D projects - Risk capital	**5. Issues concerning Companies' Strategy** - Proving orientation knowledge and long-term visions - Technology assessment - Technology council - To increase awareness of the importance of innovations
3. Other Infrastructure plus Technology Transfer via - Technology centres - Information and consulting - Demonstration centres - Cooperation, networks, personnel (e.g. change of job)	**6. Education and Training** Early arrangement of training opportunities for potential needs **7. Regulative Policy Measures** - Competition policies - Legal framework - Influence on private demand

Source: Meyer-Krahmer/Kuntze 1992

A remarkable **change and expansion** of the R&D infrastructure took place with the **unification of Germany** in 1990. Following an initial painful process of shrinkage, the research capacities of the new Länder now constitute a generally compatible part of the total German research landscape (see Chapter 7 of this study), and are now also beginning to participate in EC cooperations.

2.2 RTD Policy of the European Community

The Single European Act (ratified in 1987), which first formulated a European RTD policy, gave this policy equal status with other competences of the European Community such as economic, social and competition policy (cf. Chapter 1). Its **most important aim** was the **strengthening of the international competitiveness** of European industry in worldwide competition, especially with the USA and Japan, in sectors with a high technological content. Central to EC RTD support are the high tech sectors such as information and communication technologies, biosciences and materials research. The RTD support of the EC is primarily oriented towards projects which can be carried out more rationally and more efficiently at a European level, and from which genuine value added due to transnational cooperation can be expected.

The main instrument for EC RTD policy is the **"Framework Programme of Community Activities in the Field of Research and Technological Development".** Practical realisation of the aims of the Framework Programme takes place in so-called "specific" programmes which describe in detail the scientific topics and procedures for carrying out the programmes. The First Framework Programme (1984 to 1987) was followed by the Second Framework Programme (1987 to 1991) and the Third Framework Programme (1990 to 1994), which establishes three focal areas (cf. Figure 2.2.1): basic technologies, management of natural resources and management of intellectual resources. These in turn are broken down into six guidelines and a series of specific programmes:

1) Information and communications technologies (e.g. ESPRIT/RACE/Telematics),
2) Industrial and materials technologies (e.g. BRITE/EURAM/BCR),
3) Environment (e.g. Environment/MAST/Marine technologies),
4) Life sciences and technologies (e.g. Biotechnology, ECLAIR, FLAIR),
5) Energy (e.g. JOULE Net, ITER, Jet),
6) Human capital and mobility.

Financially speaking the Third Framework Programme, with a volume of 5,700 MECU and just under 5% of the EC budget, is relatively modest in proportion to the other tasks of the EC. Regarded in absolute terms, however, EC RTD support is becoming increasingly interesting, especially in view of the expected decreases in the flow of RTD funding from national sources. The growing importance of the EC in the field of science and technology becomes even more apparent if one takes into account the substantial efforts that have been made since the late 1980s with regard to the structural fund (regional fund, social fund, agricultural fund) to strengthen the research and technology base particularly of the less developed regions of the EC.

Figure 2.2.1: Overview over the Third and Second Framework

Framework Programme 1990–1994

Focal areas	Sums in million ECU	Proportion of total budget (%)
I. Enabling technologies		
1. Information and communications technologies	2 221	38.9
— Information technologies	1 352	
— Telecommunications	489	
— Development of technological systems of general interest	380	
2. Industrial and materials technologies	888	15.6
— Industrial and materials technologies	748	
— Measurement and testing	140	
II. Management of natural resources		
3. Environment	518	9.1
— Environment	414	
— Marine science and technology	104	
4. Life sciences and technologies	741	13.0
— Biotechnology	164	
— Agricultural and agro-industrial research (incl. fisheries)	333	
— Biomedical and health research	133	
— Life sciences and technologies for developing countries	111	
5. Energy	814	14.3
— Non-nuclear energies	157	
— Nuclear fission safety	199	
— Controlled thermonuclear fusion	458	
III. Management of intellectual resources		
6. Human capital and mobility	518	9.1
Total[1]	**5 700**	**100.0**

[1] including ECU 57 million for the centralised management of the dissemination and exploitation of research results, and ECU 550 million for the Joint Research Centre (JRC)

Framework Programme 1987–1991

Focal areas	Sums in million ECU	Proportion of total budget (%)
1. Quality of life	375	6.9
1.1 Health	80	
1.2 Radiation protection	34	
1.3 Environment	261	
2. Towards a large market and an information and communications society	2 275	42,3
2.1 Information technologies	1 600	
2.2 Telecommunications	550	
2.3 New services of common interest (including transport)	125	
3. Modernization of industrial sectors	845	15,6
3.1 Science and technology for the manufacturing industry	400	
3.2 Science and technology of advanced materials	220	
3.3 Raw materials and recycling	45	
3.4 Technical standards, measurement methods and reference materials	180	
4. Exploitation and optimum use of biological resources	280	5,2
4.1 Biotechnology	120	
4.2 Agro-industrial technologies	105	
4.3 Competitiveness of agriculture and management of agricultural resources	55	
5. Energy	1 173	21,7
5.1 Fission: nuclear safety	440	
5.2 Controlled thermonuclear fusion	611	
5.3 Non-nuclear energies and rational use of energy	122	
6. Science and technology for development	80	1,5
7. Exploitation of the sea bed and use of marine resources	80	1.5
7.1 Marine science and technology	50	
7.2 Fisheries	30	
8. Improvement of European S/T cooperation	288	5,3
8.1 Stimulation, enhancement and use of human resources	180	
8.2 Use of major installations	30	
8.3 Forecasting and assessment and other back-up measures (including statistics)	23	
8.4 Dissemination and utilization of S/T research results	55	
Total	**5 396**	**100,0**

Source: Kommission der EG 1990 b

For the "Framework Programme" the basic rule applies that EC support funding should be awarded **without regard to national proportional representation** or even quotas. The success rate in the approval of project applications is influenced by several different factors in the individual programmes, but mainly by the number and quality of applicants: the **number of applications** has **risen** substantially in the past few years, with the result that the acceptance rates have dropped continuously - in information technologies (ESPRIT II) the approval rate in the selection following the last call for proposals was only about 20%.

EC RTD support is primarily in the form of **contracted research with cost sharing**. One of the most important formal prerequisites for the approval of a project proposal is proof of transnational cooperation with partners from the EC. Participation in a "consortium" is open to every enterprise, R&D institute and establishment of a higher education institute in the twelve member states (it is now also possible for partners from the EFTA countries, Central European and Eastern European countries to participate in EC programmes). The selection of projects for all programmes is based on the following general criteria (cf. Kommission der EG 1990b):

- compatibility of the proposal with the content and aims of the relevant programme;
- precompetitive character of the proposed R&D activities;
- transnationality of the project;
- scientific and technical quality and originality of the project proposal;
- European dimension of the proposal: value added through European cooperation that could not be attained at a purely national level;
- transparent finance plan; proof of technical competence of the proposer; provision of effective programme management;
- technical and economic usefulness;
- exploitation perspectives of the expected results;
- composition of the partnership.

In addition, there are specific criteria for the individual programmes (e.g. participation of SMEs, contribution to cohesion, etc.), and these vary according to the stated objectives. The decision on the proposals is usually taken in a three-stage procedure: with the help of independent referees, a list of proposals that are most worthy of support (based on technical and scientific aspects) are selected. This list of selected projects is then passed on to the appropriate programme committee for their appraisal. The final decision is made by the appropriate Directorate General of the Commission.

2.3 Interfaces between German and Community Policy

2.3.1 Strategic Role of EC RTD Policy from the German Perspective

The starting point for an analysis of the strategic role of EC RTD policy for German RTD policy must consist in establishing its relative importance compared with national efforts in the field of research and technology. The major share of the total budget for research and technology in the Federal Republic of Germany is, as already indicated, accounted for by industry (1981: 55%, 1992: 59%); the Federal Government is the second largest financier (1981: 26%, 1992: 22%); in third place come the Länder governments (1981: 17%, 1992: 16%); measured on this scale, the proportion coming from abroad - which in these statistics includes the **Commission of the European Community (CEC)** - is **only a very small proportion of research financing** (1981: 0.9%, 1992: 2.4%). However, it is remarkable that in the course of the last decade the financing share of the Federal Government has decreased, whereas the share of industry has grown and the share from sources abroad (including the EC) has **more than doubled**. Figure 2.1.1 in section 2.1 shows - in a simplified form and sometimes as estimates - the share of EC financing in the RTD expenditure of the most important actors in the German research and innovation system. This diagram clearly demonstrates the relatively low importance of EC financing.[11]

According to statistics of the BMFT, in the period 1987-91 a total of 1,308 thousand million DM (652,940 million ECU) was received by German actors from DG XII and DG XIII. Compared with the total German domestic expenditure on R&D in the same period (327.5 thousand million DM; BMFT 1993), EC funding represented 0.4% of the financing of the R&D budget in Germany. Relative to expenditure by the Federal Government (1987-91: 72.4 thousand million DM), EC funding received by Germany represented an approximate 1.8% share, and relative to civil, direct project support by the Federal Government (21.9 thousand million DM, not including expenditure by the Federal Ministry of Defence (BMVg)) it represented ca. 5.9%.

---

11 As reliable data on the EC funding for the individual actors were not available, they had to be estimated from various sources. The EC contribution for industry (ca. 1.000 million DM) and for the HEIs (ca. 170 million DM) refer to the contracts signed with the EC in 1991; these figures are certainly too high, as the end of Framework Programme II and the beginning of Framework Programme III overlapped. The data for the MPG is based on the average annual EC contribution 1987-91 (annually ca. 5.6 million DM) and for the FhG 1985-90 (annually ca. 8.6 million DM). Figure 2.1.1 can only approximately show the importance of the EC contribution.

Figure 2.3.1.1: R&D Expenditure of Germany and the EC according to Sections of the EC Framework Programme in % of the Expenditure of Germany/ the EC 1987 -1991 and 1992 - 1993

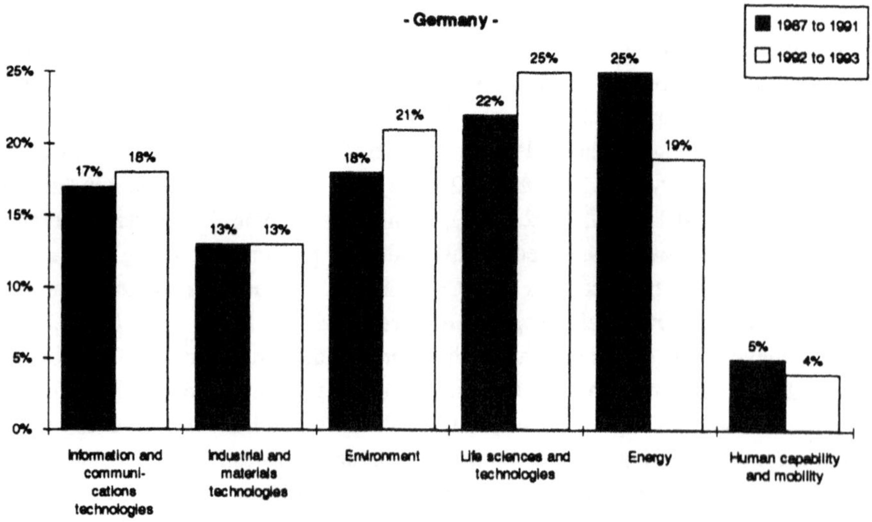

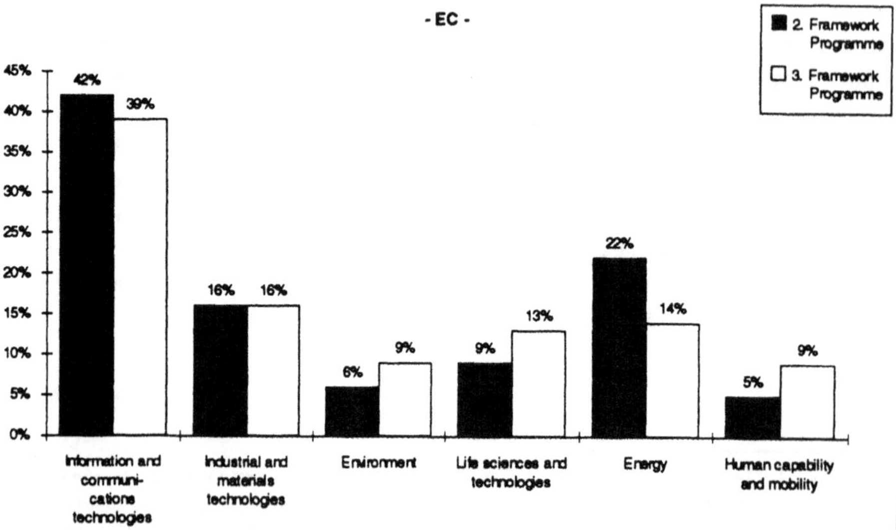

Source: BMFT, FhG-ISI

However, this should not hide the fact that in some fields of research and technology EC financing has gained considerable significance in Germany. A **comparison** of RTD efforts by the **German Federal Government and the CEC according to main research topics** clearly shows where EC financing is an influential factor: figure 2.3.1.1 gives a percentage breakdown of the support activities of the German Federal Government and the CEC according to the topics of the Framework Programmes. The Federal Government is clearly more active than the EC in the fields of energy, life sciences and technologies (e.g. biotechnology) and environment; in the area of industrial and materials technology their activities are roughly equivalent. In the field of **information and communications technologies**, however, the **EC is relatively much more active** than the Federal Government; in absolute terms, EC financing of R&D activities in the field of information technology in Germany is equivalent to approximately **one fifth to one quarter of R&D support expenditure by the Federal Government** in this field (statements from interviews).

In the RTD support of the EC, it is the quality of the supported projects, measured according to the criteria listed above, that is meant to be decisive, and not the principle of "juste retour"(returns of funds paid in to the EC). With regard to Germany, this "juste retour" is not "poor" at present (as is sometimes claimed in Germany), but in fact lies only slightly below the share of Germany in the financing of the EC (cf. KoWi 1992; the statistics given relate to DG XII).

Taking as an example the contracts signed by the CEC in 1991, shown in Table 2.3.1.1, German participants receive 22% of available EC funding; this means that Germany is **satisfactorily represented**. However, there are important deviations from this mean value: the funding of the predominantly industrially-oriented programmes of DG XIII is mainly received by project participants in German industry, whereas much less of it comes to higher education institutes and non-university R&D institutes; even in the relatively scientifically oriented programmes of DG XII, contractors from industry predominate slightly over HEIs and RDIs measured in terms of the volume of funding. If these values are compared with the average for all EC countries, then it is the **under-representation of the German HEIs** that is particularly obvious.

From time to time, German R&D institutions and their organisations point out various **problems and weaknesses of EC RTD support** and its procedures. On closer examination, it is clear that some of these complaints are to be taken very seriously, and others less seriously. These problem areas will be taken up again in the empirical analyses in Chapters 4, 5 and 6 of this study, but we should like at this point to give the following indications (cf. also KoWi 1992):

Table 2.3.1.1: German Share of EC Promotion Funds (Contracts signed 1991)

	EC total	German Share							
		total		HEIs		R&D Institutes		Industry and others	
	MECU[1)]	MECU	% of Col. 2	MECU	% of Col. 3	MECU	% of Col. 3	MECU	% of Col. 3
1	2	3	4	5	6	7	8	9	10
GD XII[2)]	508.7	93.6	18.9	27.6	28.7	33.2	34.4	35.5	36.9
GD XIII[3)]	2661.1	608.4	22.8	56.4	9.3	76.8	12.6	475.2	78.1
Total	3174.8	704.7	22.2	84.0	11.9	110.0	15.6	510.7	72.5

For comparison: Share of Various Actors in all EC Countries

	EC total		HEIs	R&D Institutes	Industry and others
GD XII	508.7		ca. 32%	ca. 38%	ca. 30%
GD XIII	2661.1		ca. 12%	ca 12%	ca. 76%

1) MECU = Millions ECU; Exchange rate 22.07.1992 = 2.03862

2) Source: GD XII report

3) Source: Calculated on basis of GD XIII: List of Projects Financed 1991

Source of Table: Report of Working Group BMfT and the Conference of Education and Arts Ministers on Share of EC Research Promotion Funds for German Science, dated 15 April 1993

- Scientists at higher education institutes in particular complain that some of their research fields are not adequately represented in the **topics of EC RTD support**. However, here they have failed to appreciate the character and tasks of EC RTD support, which - at any rate until now - has not been intended as general research support, but pursues explicit aims such as increasing the competitiveness of European industry by improving its technological competence, and performing research tasks which could not be carried out at a purely national level.

- Occasionally the suspicion is voiced that research projects supported by the EC may not have sufficiently high **scientific standards**. Anticipating the results of our empirical survey, we can state here that this accusation has virtually ceased to be heard since the Second Framework Programme - on the contrary, it is sometimes regarded as a distinction among German participants to take part in projects supported by the EC in some fields.

- There are complaints about receiving **information** too **late** and about **application times** being too **short**. This is indeed a serious problem, which can often only be overcome by potential participants trying informally to obtain information about the planned focal topics of a programme while it is still in the process of formulation. Expert coordinators (suggested by the BMFT) and, since the start of Framework Programme III, the newly-formed Liaison Office of the German Scientific Research Organisations (KoWi) have been attempting to provide assistance on this aspect by passing on information to potential German participants in the research system at the earliest possible time.

- Complaints about the **excessive amount of bureaucracy** involved **in the application procedure** and the **management of the project** do not take into account the fact that in international project cooperations with many different partners, coordination and control are always more complex than, for example, in the running of an individual project supported by the German Research Association (Deutsche Forschungsgemeinschaft, DFG).

- Complaints about the dominant role of the **English language** in particular in the running of EC projects fail to recognise that, for a long time now, English has been the usual language for international scientific communication; additionally, the English language is required to communicate in the transnational project consortia.

- Disappointments caused by the **high rejection rates** for project proposals are frequently justified. In some fields, only 20% of the proposals submitted can be approved. However, this is a consequence of the recent rapid growth of interest generated throughout Europe in participation in EC RTD support, with which the substantial growth of EC funding cannot keep pace. The CEC cannot be blamed for this; it asked for more funds than were granted to it by the Council of Ministers. Nevertheless, low rates of approval do represent a waste of resources, and thus constitute a (supra)national economic problem, if the expenditure for application is high.

- **Contract research institutions**, in particular the Fraunhofer Society (FhG) have problems with the EC support model, in which projects are on a cost sharing basis ("cost shared actions"). Industrial partners of Fraunhofer institutes are sometimes prepared to take over half of the FhG's contribution (i.e. 25% of expenses) in joint projects which are market-oriented, but are unwilling to do so in projects with a higher risk factor which have a more forward-looking orientation, or in projects in the field of preventive research. In projects of this kind the FhG institutes thus have to resort to its basic funding, which was neither originally intended for this purpose, nor is sufficient in volume to meet it. This situation causes considerable difficulties in the participation of Fraunhofer institutes in EC RTD programmes.

- German **higher education institutes** often do not yet have an adequate apparatus for converting results into **patents and licences**. However the Commission sets great store by the use and exploitation of knowledge gained from EC research projects. Here the German HEIs are at a competitive disadvantage; their situation in this respect is roughly comparable to the British higher education institutes.

Despite these problems it can be assumed that the **interest** of German R&D institutions in participation **in EC RTD support** will **continue to increase**. One important reason for this is that German R&D institutions are becoming increasingly aware of the growing importance of international - and in this context, European - cooperation in the field of research and development. Another important reason is that the "years of abundance" in German national RTD support are over and that, in view of this, German R&D institutions in their search for funding are attempting to compensate the reductions in national funding and to increase their financing by using the possibilities of EC RTD support.

2.3.2 The Shaping and Influencing of Community Policy from the German Viewpoint

The German attitude towards EC RTD policy may **now** be characterised as **"constructive pragmatism"**. Although the subsidiarity principle is still the main criterion in official positions (EC support only if "European value added" can be expected or if an R&D topic cannot be tackled at a purely national level; cf. BMFT 1993), nevertheless the actual practical division of work between Bonn and Brussels tends more to follow an incremental pattern: if, for instance, the areas supported in the field of information technology in BMFT programmes and in EC programmes are examined more closely, a clear division of work according to rigid principles will be difficult to find.

The form and content of the **participation procedures** for national bodies in shaping framework programmes are regulated as follows: The EC Council of Ministers makes its decision on the Commission's proposal for the design of an EC Framework Programme after hearing the European Parliament and the Economic and Social Committee. The Council's decision has to be unanimous and alterations to the CEC's proposal can only be made on a basis of unanimity.[12]

Before the decision by the Council of Ministers, the Commission develops the **Framework Programme** with the assistance of numerous experts from science and industry. The Commission also obtains statements from three own **advisory committees:** the Committee for the European Development of Science and Technology (CODEST), the Industrial Research and Development Advisory Committee (IRDAC), and the Scientific and Technical Research Committee (CREST). In CREST, each national government is represented by two members; from the German side CREST includes representatives from the German Federal Ministry for Research and Technology (BMFT) and the Federal Ministry for Economic Affairs (BMWi); the 16 Länder have a representative with observer status. For the development of the**"specific programmes"** of the Framework programme, **special ad hoc support committees** are formed, in which again the Member Countries are generally represented by two experts.

Formally speaking, the formation of an opinion on European RTD policy within the German Federal Government takes place via the usual processes of the joint rules of order of the cabinet ("Gemeinsame Geschäftsordnung") and - in accordance with the **principles of federalism** - with the participation and in consideration of the opinions of the

12 With the coming into force of the Treaty of European Union (Maastricht Treaty), this will have changed: the European Parliament will then have a right of veto, and planning and decision on the Framework Programmes will be according to the so-called "co-decision" process.

Federal Länder. According to a special agreement with the Länder,[13] representatives of theirs regularly participate in the special programme committees of the specific programmes, in the CREST and in discussions preparatory to sessions of the EC Research Council.

Analogous to the drafting process of national programmes, another particularly important element in the formation of government opinion in Germany on EC RTD policy is the **dialogue of the BMFT with the large scientific organisations** (the German Research Association, the Max Planck Society, the Fraunhofer Society and others), the leading organisations of industry and the trade unions; there are also discussions with the Federal States Commission for Educational Planning (Bund-Länder Kommission zur Bildungsplanung und Forschungsförderung, BLK). These expert discussions are intended to round off the formation of opinion.

2.3.3 Interaction with the Addressees of EC RTD Policy

Only in the last few years has the interaction between the German addressees of EC RTD policy and those responsible at the CEC Commission **grown** in scope and intensity. Apart from the official committees described in section 2.3.2, active "interference" from German sources during preparations for the First Framework Programme and during the running of the programme came mainly from a few large German enterprises, especially in the field of information technology. Small and medium-sized firms and scientific institutions took virtually no part in active communication with the Commission, nor did they receive much support at that time from the policy administration organisations of German RTD policy.

In the meantime, the situation has changed considerably. Both on the side of the EC and on the German side, there have been **considerabe efforts** to **improve the flow of communication** between the addressees of EC RTD policy and the CEC. These include:

- **EC advisory offices for enterprises**: there are now more than 35 EuroInfo Centres throughout Germany. In addition, there are 80 contact partners in the EC ad-

13 "Vereinbarung zwischen der Bundesregierung und den Regierungen der Länder über die Unterrichtung und Beteiligung des Bundesrates und der Länder bei Vorhaben im Rahmen der Europäischen Gemeinschaften in Ausführung von Artikel 2 des Gesetzes vom 19. Dezember 1986 zur Einheitlichen Europäischen Akte vom 28. Februar 1986" of December 17, 1987

visory offices of chambers of industry and commerce; the EC contact partners in the German sectoral industrial associations should also be mentioned.

- More and more **Länder** have now recognised that it is important actively to support their own science and industry in participation in EC RTD support. Several Federal Länder now have their own **representatives in Brussels,** and have founded departments for EC affairs and/or EC liaison offices. There is also a growing number of **EC experts** at **higher education institutes**, who give information about support possibilities and help scientists to submit proposals.

- In order to provide expert advice on specialist fields and programmes for interested parties in Germany - and especially for those having few contacts with supranational organisations, such as SMEs and scientists from small institutions - the Federal Minister for Research and Technology has initiated a **system of specialist contact partners**. These expert coordinators are **usually also responsible for the administration of national support measures in relevant specialist fields**, and thus have comparatively the best overview of national and EC support possibilities within one specialist field. These mediators are usually institutionally affiliated to the BMFT or to the national research centres, or to the project administrator organisations of the Federal Government.

- The **Liaison Office of the German Scientific Research Organisations (KoWi)** is gaining importance; this was initiated in 1991 by the Alexander von Humboldt Foundation, the Association of National Research Centres, the German Academic Exchange Service, the German Research Association, the Fraunhofer Society, the Conference of Principals and University Rectors, the Max Planck Society and the Donors' Association for the Advancement of Science in Germany. The task of the KoWi is to support and promote scientific cooperation in Europe, and particularly to improve the use made of EC support offers by German scientists. For this purpose the KoWi has two offices in Brussels and in Bonn. The KoWi regularly produces and distributes informative publications and organises informative events.

- As an EC Commission measure, **decentralised relay centres for research and technology** (EG-Verbindungsbüros für Forschung und Technologie) were established, which are intended principally to facilitate access to EC RTD programmes for SMEs and to improve dissemination of project results; in Germany centres were opened in Berlin, Leipzig, Mühlheim/Ruhr and Stuttgart in 1993/94 .

A considerable number of **new addressees of EC RTD policy** is forming at present **in the New Länder**. Until now, the dramatic internal re-organisation process in Eastern Germany following German unification in 1990 has slowed down active participation of R&D institutions from the New Länder (see also Chapter 7). In the meantime, however, a research infrastructure of high quality - though markedly reduced in size compared with the previous GDR infrastructure - has formed, which is eager to participate actively in the EC R&D community.

3. The Structure of German Participants in the Second Framework Programme

Precise and systematically coordinated data on the structure of German participants in the research and technological development support of the European Community under the Second Framework Programme were not publicly available until now. In order to fill this information gap, ISI had access to the AMPERE data bank of the Commission of the European Community, which placed these data on German participation at our disposal. This enabled us to undertake a **structural analysis of German participants** according to type of institution (enterprise, higher education institute, non-university R&D institute), size, position in the project consortium, regional distribution and EC programme affiliation. The analysis was mainly confined to the **frequency of participation in EC projects**; only in a few cases could the financial contribution of the EC be used as an indicator. Unfortunately, for some programmes information is only available on the EC's financial contribution to the EC project as a whole, and not on the shares of the German participant.

A comparison of the participation structure with other EC countries will be undertaken in a "synthesis report" by the French institute AXION after completion of all the IMPACT studies.

3.1 Data Base

Based on the AMPERE data bank of the CEC, a data bank was set up by ISI containing data on all the German participants in projects of the DG XII and DG XIII within the RTD support of the Framework Programme II. A considerable amount of processing work had to be done on the AMPERE data, as in some cases there was no classification according to type of institution, and many participants were not classified correctly. Therefore the information about each German participant was classified with the help of the "Hoppenstedt" directory for enterprises or the "Vademecum der deutschen Lehr- und Forschungsstätten".[14] Some examples of incorrect typifications in data provided by the EC Commission are: confusion of large and small enterprises, the assigning of R&D laboratories of large enterprises to "non-university R&D institutions" and the classifica-

---

14 The typification of 48 institutions (from DG XII projects) had to be taken over without confirmation, as they could not be checked; 47 institutions (from DG XIII projects) could not be identified from the information available, and were therefore not assigned to a group.

tion of institutes associated with universities in the category "higher education institutes", although they are not subject to direct university administration.

Categorisation was undertaken by ISI according to the following manner:

- **BIGs (BIGs):** industrial companies with 500 and more employees, their R&D laboratories and their R&D centres;

- **SMEs:** small and medium-sized enterprises with less than 500 employees;

- **Higher education institutes (HEIs):** departments or institutes in higher education institutes, technical higher education institutes and technical colleges ("Fachhochschulen") insofar as they are subject to direct university/technical college administration;[15]

- **Non-university R&D institutes (RDIs):** institutes of the Fraunhofer Society (FhG) and the Max Planck Society (MPG), national research centres (GFE), institutions on the "blue list" (BLE), laboratories of the AIF, state and federal research establishments, laboratories associated with higher education institutes (insofar as they are not administratively affiliated to a university), transfer centres, private non-profit R&D institutes, research-intensive consultancy and engineering bureaux, other institutions such as e.g. the TÜV (Association for Technical Inspection), museums and registered societies.

3.2 Structure of Participants according to Types of Institutions

Altogether, German institutions participated 2,108 times in the R&D projects supported by the EC (subsequently referred to as EC projects) within the Second Framework Programme.[16] As can be seen in Table 3.2.1, higher education institutes (HEIs) with 32.2% and large companies (BIGs) with 31.3% participated almost equally often. By comparison, the proportion of small and medium-sized firms (SMEs) participating is remarkably low. German SMEs only participated 166 times in EC projects (7.9% of all German

15 The question of administration is very important in this context: the EC categorises German institutes associated with universities ("An-Institute") as university institutions, although they are not directly administered by the universities, and according to the Frascati Manual (OECD 1992) should therefore be considered as non-university R&D institutions; for this reason, we have placed these associated institutes in the "RDI" group.

16 German establishments took part in DG XII projects 1127 times and in DG XIII projects 981 times.

participations) (cf. Table 3.2.1). If the participation of German SMEs in the programmes of the Second Framework Programme (7.9%) is compared with the participation from Member Countries (15.4%) (CREST 1992, p.14), German SMEs are clearly under-represented. [17]

Table 3.2.1: Typology of German Participants in the Second EC Framework Programme for RTD (1987-91)

Type of organisation	EC programmes except I&C		I&C programmes of the EC[1)]		Second Framework Programme total	
	absolute	%	absolute	%	absolute	%
BIGs	246	21.8	414	42.2	660	31.3
SMEs	75	6.7	91	9.3	166	7.9
HEIs	442	39.2	237	24.2	679	32.2
RDIs	364	32.3	192	19.6	556	26.4
unknown			47	4.8	47	2.2
Total	1127	100.0	981	100.0	2108	100.0

BIGs: Large firms with 500 and more employees
SMEs: Small and medium-sized firms with less than 500 employees
HEIs: Higher education institutes
RDIs: Non-university R&D institutes
1): ESPRIT, RACE, DRIVE, DELTA, AIM, EUROTRA, VALUE

If participation in EC programmes is measured by the **amount of funding** received by German project participants from the CEC, there is a shift in participation structure towards the enterprises, especially the large ones (cf. Table 3.2.2).[18] Taking the criterion of the EC Commission's financial contribution, the share of the HEIs becomes appreciably smaller, as these receive less funding on average, reflecting the division of tasks within the projects.

17 In the European Communities as a whole, 15.4% SMEs, 15.4% large enterprises, 34.6% higher education institutes, 31.4% non-university R&D institutes and 3.2% other institutions are taking part in the Second Framework Programme (CREST 1992, p.14).

18 Since information on funding given to German participants is only available for DG XII RTD programmes, no statements can be made on funding for the Second Framework Programme as a whole.

Table 3.2.2: Financial Contribution of DG XII to German Participants in the Second Framework Programme

Type of organisation	Number of projects (absolute/ in %)	Mean value of EC financial contribution (in ECU)	EC financial contribution (in Mio. ECU/in %)
BIGs	246 / 21.8%	220746	54.3 (30.9%)
SMEs	75 / 6.7%	189559	14.2 (8.1%)
HEIs	442 / 39.2%	116193	51.4 (29.3%)
RDIs	364 / 32.3%	152720	55.6 (31.7%)
Total	1127/100.0%	155694	175.5 (100.0%)

Non-university R&D institutes receive the greatest share of funding (31.7%), followed by BIGs (30.9%) and HEIs (29.3%). In terms of funding received, the share of SMEs goes up very slightly to 8.1 %, underlining their low participation rate relative to other types of participants (cf. Table 3.2.2). This corresponds to their role in German RTD programmes: for comparison, here are statistics on the share of funding received by SMEs from the German Federal Ministry for Research and Technology (BMFT) in the form of project support related to specialised programmes: altogether, from 1987 to 1991, SMEs in the Federal Republic of Germany received 1,199.3 million DM from the BMFT in the form of direct project support (BMFT 1990 and 1993, own calculations); since during this period the BMFT spent a total of 15,832.7 million DM on direct project support, the share of SMEs in total funding was 7.6%.[19] The share of SMEs in support related to specialised programmes of the BMFT (7.6%) is thus slightly lower than the participation rate of German SMEs in the RTD programmes of DG XII (8.1%) measured in terms of the share of funding.

However, the direct project support only has a 44% share of the BMFT support of SMEs which amounts from 1987-91 to a total of 2,486.8 million DM and includes other policy measures.

Numerous enterprises, HEIs and RDIs participated several times in the EC programmes. If the multiple participation figures are reduced in each case to the large enterprise, the higher education institute or the research organisation to which an R&D institute is

19 If the BMWi's funding of joint industrial research (622,7 million DM in 1987-91) is included and related to the R&D expenditure of the BMWi and the BMFT on direct project promotion (totalling 18,553.6 million DM in 1987-91), the share of funding received by SMEs goes up to 9.8% (BMFT 1990; own calculations).

affiliated, the basic population of participants is reduced.[20] If these figures are related to the frequency of participation of the various types of institutions, a measurement is obtained which expresses, so to speak, the **"density of participation in EC projects"** (cf. Table 3.2.3). German participation in EC support programmes can thus be reduced to certain large enterprises, HEIs and RDIs which frequently participate in EC projects. On average, each university participates in 9.8 EC projects, and the non-university R&D organisations and their institutes participate in an average of 4.6 EC projects. Whereas on average SMEs only participate in approximately one EC project (ratio 1:1.3), the "EC participation density" of BIGs (ratio 1:4.4) is much more higher (cf. Table 3.2.3).

Table 3.2.3: Frequency of German Participation in the Second Framework Programme

Type of organisation	Total participation	Frequency of participation	Ratio
BIGs	151	660	1:4.4
SMEs	132	166	1:1.3
HEIs	69	679	1:9.8
RDIs	122	556	1:4.6
unknown	43	47	1:1.1
Total	517	2108	1:4.1

The thesis often put forward, that German participants in EC RTD support basically consist of a small community of actors which - with the exception of SMEs - participate very often in EC projects, is substantiated if the participation frequency of the **"Top Ten" national actors** in each category is considered. In the case of the BIGs (BIGs), the higher education institutes (HEIs) and the non-university R&D institutes (RDIs), between 43% and 61% of all participations in the relevant category are accounted for by the 10 institutions that are the most frequent participants (cf. Table 3.2.4). Top of the participation league in the BIGs is Siemens AG, with 95 participations in EC programmes within the Second Framework Programme; in the HEI category, RWTH

20 The basic population of participants goes down if multiple participation of the R&D institutions is handled according to the following rule:
1. The various groups of large firms are put together under the name of the concern. This also applies to daughter companies, if the name of the concern appears in the firm name of the daughter company (e.g. AEG Olympia -> AEG).
2. The individual Fraunhofer (FhG) and Max Planck (MPG) institutes are grouped together under the Fraunhofer Society and the Max Planck Society. The reason for doing this is to make these institutes comparable to university institutes, which may be relatively independent within the university or may be organisationally tied, but are not entered in the data banks under the name of the institute but in most cases as "university X".

Aachen (the Rhineland-Westphalia Technological University) tops the list with 44 participations, and among the RDIs the institutes of the FhG, with 85 participations. **German participants in the Second Framework Programme obviously form a definite "community"** which, once the circle of project consortia has been entered, take part repeatedly in various different project consortia. From the interviews with non-participating SMEs, it emerges clearly that the difficulties experienced by SMEs in relation to participation in EC projects consist above all in gaining an entry to the "network" (cf. Chapter 4).

Table 3.2.4: The "Top Ten" of German Participants in the Second Framework Programme

BIGs	N	%	HEIs	N	%	RDIs	N	%
1. Siemens	95	14.4	TH Aachen	44	6.5	FhG	85	15.3
2. SNI	49	7.4	Uni Stuttgart	39	5.7	MPG	76	13.7
3. SEL	42	6.4	TU München	33	4.9	GSF	39	7.0
4. Daimler	30	4.5	Uni Karlsruhe	31	4.6	DLR	30	5.4
5. AEG	26	3.9	TU Berlin	29	4.3	GMD	28	5.0
6. Dornier	24	3.6	Uni Hamburg	25	3.7	KfK	23	4.1
7. Philips	24	3.6	Uni Bochum	23	3.4	FZ Jülich	20	3.6
8. Krupp	22	3.3	Uni Hannover	23	3.4	BIBA	12	2.2
9. Bosch	21	3.2	Uni München	23	3.4	Wi. und Inf.	11	2.0
10. MBB	18	2.7	TH Darmstadt	21	3.1	GFAI	10	1.8
						PTB	10	1.8
Total	351	53.2	**Total**	291	42.9	**Total**	344	61.9
Total BIGs	660	100.0	**Total HEIs**	679	100.0	**Total RDIs**	556	100.0

3.3 Structure according to Participants' Position in Project Consortium

Before their application is submitted, prospective project participants must have already reached an agreement on their respective roles in the consortium. In EC projects different rights and duties are attached to different positions within the consortium (Kommission der EG 1990b, p. 37ff). The CEC distinguishes four functions:

1. **Main contractors** are responsible for project coordination and represent their partners to the Commission. They are responsible signatories for the implementation of the contract and have decisive importance in the management of the pro-

ject, as subsidies for all the participants are transferred by the EC to the main contractor. They are also responsible for seeing that the time schedule is adhered to. These tasks of project coordination require experience in the management of supranational projects and necessitate appropriate and adequate organisational, financial and technical resources.

2. **Contractors** participate in carrying out and financing the project; they are co-signatories of the contract with the EC and assume full responsibility. They are therefore fully entitled to exploit the project results.

3. **Associates** participate in carrying out and financing the project; as they are not contract signatories with the EC and do not assume full responsibility; they only have limited access to project results.

4. **Subcontractors** provide (technical) services for partners in a consortium; they are directly contracted to the consortium partner in each case and are not contracted to the CEC. They have no rights to the results of the project.

In 56.8 % of cases, German participants assume the role as **partners** in EC projects. In 28.3 % of cases they assume the role of **main contractor**, whereas only in less than 10 % of cases do they participate as associates (cf. Table 3.3.1).[21]

When the figures are broken down according to type of R&D establishment, it becomes apparent that the group of SMEs, with their limited R&D capacities and in-house resources, are not very important as leaders of EC projects. In 203 of 596 cases (34.1 %) where consortia had German project leaders, the role of main contractor was taken on by a RDI; 30.9 % and 28.7 % of main contractors respectively were HEIs or BIGs. In the category of **associates**, there is above-average representation of the higher education institutes, at 40.3 %.

21 Subcontractors are not systematically recorded in the AMPERE data bank, as there is no contractual agreement between subcontractors and the EC. Thus data on subcontractors do not represent the actual number of subcontractors, but only the number which (by chance) have been recorded in the data bank.

Table 3.3.1: Participants according to Position in Consortium (Second Framework Programme)						
Number function % Type of partici-pation %	BIGs	SMEs	HEIs	RDIs	Other/un-known	Total
Associate	34.0 16.5 5.2	25.0 12.1 15.1	83.0 40.3 12.2	52.0 25.2 9.4	12.0 5.8 25.5	206.0 100.0 9.8
Main contractor	171.0 28.7 25.9	34.0 5.7 20.5	184.0 30.9 27.1	203.0 34.1 36.5	4.0 0.7 8.5	596.0 100.0 28.3
Partner	442.0 36.9 67.0	102.0 8.5 61.4	354.0 29.5 52.1	270.0 22.5 48.6	30.0 2.5 63.8	1198.0 100.0 56.8
Sub-contractor	12.0 11.7 1.8	5.0 4.9 3.0	55.0 53.4 8.1	30.0 29.1 5.4	1.0 1.0 2.1	103.0 100.0 4.9
Other/unknown	1.0 20.0 0.2		3.0 60.0 0.4	1.0 20.0 0.2		5.0 100.0 0.2
Total	660.0 31.3 100.0	166.0 7.9 100.0	679.0 32.2 100.0	556.0 26.4 100.0	47.0 2.2 100.0	2108.0 100.0 100.0

3.4 Participation in Specific Programmes

German R&D institutions participate most frequently in the ESPRIT(II) Programme (31.6%), followed by the BRITE/EURAM Programme (15.1%), SCIENCE (10.7%), JOULE (7.3%) and RACE (6.7%). **German participation is mainly concentrated within these five support programmes**; taken together, they account for almost three-quarters (71.4%) of participations from the Federal Republic of Germany (cf. Table 3.4.1). Of the (eight) programme lines considered here, almost half of the German participants are involved in the programme line "Information and Communication Technologies", due to the high importance assigned to the ESPRIT programme (but also to RACE and DRIVE).

Table 3.4.1: Structure of German Participants in EC Programmes

EC-Programmes (Second Framework Programmes 1987-1991)		EC Budget		German Participants	
		million ECU	%	absolute	%
1. Quality of Life		**375**	**6.9%**	**165**	**7.8%**
e.g.:	Human Gen. Analysis	15	0.3%	1	0.0%
	Medicine/Health	65	1.2%	16	0.8%
	Radiation Protection	34	0.6%	60	2.8%
	STEP/EPOCH	115	2.1%	88	4.2%
2. Information and Communication Technologies		**2275**	**42.3%**	**957**	**45.4%**
e.g.:	AIM	20	0.4%	34	1.6%
	DELTA	20	0.4%	16	0.8%
	DRIVE	60	1.1%	99	4.7%
	EURET	25	0.5%		
	ESPRIT II	1600	29.7%	667	31.6%
	RACE I	550	10.2%	141	6.7%
3. Modernisation of Industry		**845**	**15.7%**	**365**	**17.3%**
e.g.:	B.C.R	59	1.1%	①	
	BRITE/EURAM	560	10.4%	319	15.1%
	Raw Materials	45	0.8%	46	2.2%
4. Biological Resources		**280**	**5.2%**	**47**	**2.2%**
e.g.:	Agriculture	55	1.0%	③	
	BIO/BRIDGE	120	2.2%	35	1.7%
	ECLAIR	80	1.5%	11	0.5%
	FLAIR	25	0.5%	1	0.0%
5. Energy		**1173**	**21.8%**	**235**	**11.2%**
e.g.:	Decommissioning	32	0.6%	30	1.4%
	Fusion	551	10.2%		
	JOULE	122	2.3%	153	7.3%
	Radioactive Waste	80	1.5%	24	1.1%
	TELEMAN	19	0.4%	13	0.6%
6. Science and Technology for Development		**80**	**1.5%**	**42**	**2.0%**
7. Marine Resources		**80**	**1.5%**	**29**	**1.4%**
e.g.:	FAR	30	0.6%	②	
	MAST	50	0.9%	29	1.4%
8. Improvement of European S/T Cooperation		**288**	**5.3%**	**268**	**12.7**
e.g.:	DOSES	4	0.1%	③	
	EUROTRA II	7	0.1%	13	0.6%
	Large-scale facilities	30	0.6%	1	0.0%
	MONITOR	22	0.4%	10	0.5%
	SCIENCE	167	3.1%	226	10.7%
	SPES	6	0.1%	7	0.3%
	VALUE	38	0.7%	11	0.5%
Total		**5396**	**100.0%**	**2108**	**100.0%**

① Small prog. with an average project size from 2,000 to 10,000 ECU, altogether 227 German participants.

② FAR is not completed; as this sub-programme is not technologically aimed, it was not dealt with in greater detail in the study.

③ Programmes without technological aims, were not considered in more detail.

Source: Commission of the EC (1987); ISI databank of German participants in EC programmes (own calculations)

In **ESPRIT II** 38.5% of all German participants are BIGs and 27.3% are HEIs, as opposed to RDIs (18.9%) and SMEs (10%). (cf. Figure 3.4.1).[22] Almost half of all German **participations in BRITE/EURAM** are by BIGs (49.8%) and about one-fifth are by non-university R&D institutes (19.7%). By contrast, German SMEs have only a relatively small participation share in BRITE/EURAM (13.8%); this is astonishing in view of the fact that, according to experts interviewed, this programme was also oriented towards the research requirements of European SMEs. The participation share of German SMEs is even lower than the level for the EC as a whole: altogether European SMEs constitute 22% of the 1,239 participants in BRITE/EURAM (Commission of the EC 1993a, p.85). The BRITE/EURAM Evaluation Panel already considers the European participation rate for SMEs to be too low. The **reasons** mentioned by the Panel - which must surely apply to the German SMEs as well - are the limited resources of SMEs, the orientation of innovative behaviour towards developments which are close to the market, and the over-wide scope of BRITE/EURAM projects.

Due to the topics and the specific target groups it addresses, only the categories of higher education institutes (68.6% share) and non-university R&D institutes (31,4%) participate in the **SCIENCE** support programme, which ranks third in terms of German participations (cf. Figure 3.4.1). German participants in the **JOULE** programme are mainly HEIs (38.6%) and RDIs (30.1%); about one quarter (23.5%) are BIGs. **RACE I** is addressed particularly to BIGs in information and communication technologies; thus over two-thirds of German participants in this programme (67.4%) are large companies.

Figure 3.4.1: Structure of German Participants in selected EC Programmes

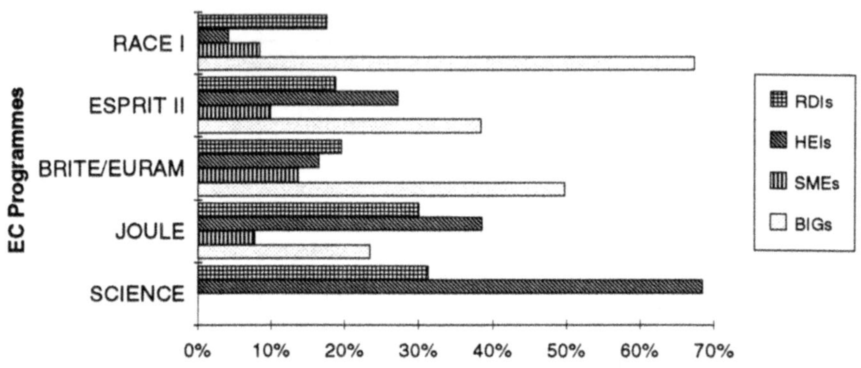

[22] For the exact percentages and the structure of German EC participants in other EC programmes see the more detailed Table A3.4.1 in the Annex.

What differences in programme participation are there between different types of participants? **BIGs** (38.9%) and **SMEs** (40.4%) both have their highest participation level of about 40% in ESPRIT II, and their second highest - about 25% - in BRITE/EURAM (cf. Table A3.4.1 in Annex). In third place for both groups comes RACE I with 14.4% of BIGs and (7.2%) of SMEs. Participation in JOULE projects ranks fourth for the latter group (7.2%), while for BIGs DRIVE projects occupy fourth place (6.4%).

ESPRIT II is the programme most frequently participated in by **higher education institutes** (26.8%) and **non-university R&D institutions** (22.7%) (cf. Table A3.4.1 in Annex). For these two groups the SCIENCE programme ranks second (22.8% and 12.8% respectively). For HEIs JOULE (8.7%) and BRITE/EURAM (7.8%) are also important. For RDIs the pattern is similar, except that the ranking of BRITE/EURAM (11.3%) and JOULE (8.3%) is reversed.

3.5 Regional Distribution

For the regional classification of German participants the Eurostat NUTS system was used, and at the level of "Regierungsbezirke" (NUTS 2) - the primary administrative divisions of the German Länder - they are comparable with other regions throughout the European Community. [23] Most EC participants in Germany are from the following regions (NUTS 2): Oberbayern (15.7% of German participants), followed by Köln (11.9%), Stuttgart (8.7%), Karlsruhe (7.7%), Darmstadt (6.0%), Berlin (5.3%) and Hamburg (4.8%) (cf. Figure 3.5.1). [24]

Regional distribution thus correponds to the **presence of research-intensive BIGs, technical higher education institutes and non-university R&D institutions** in the areas. The high proportion of EC project participants in the **area of Oberbayern**, for instance, is explained by the frequent participation of BIGs such as Siemens and MBB, the Technical University (TU) and the University of Munich, and the numerous institutes of the Fraunhofer Society and the Max Planck Society. In the administrative **area of Köln** the following participants in EC projects are situated for instance: the Rhineland-Westphalia Technological University (RWTH Aachen), large companies such as Philips, several Fraunhofer and Max Planck institutes, as well as some national research centres (e.g. the GMD, Gesellschaft für Mathematik und Datenverarbeitung, and the DLR, Deutsche Forschungsanstalt für Luft- und Raumfahrt). Frequent participants in the **area**

23 Eurostat (1992): Regions - Nomenclature of Territorial Units for Statistics (NUTS), Brussels.

24 Cf. the exact data (absolute and in %) for the regional distribution of EC participants given in Table A3.5.1 in the Annex.

Figure 3.5.1: Regional Distribution of Participants in the Second Framework Programme

of Stuttgart are the headquarters and R&D labs of Daimler-Benz, SEL and Bosch, the University of Stuttgart and several Fraunhofer and Max Planck institutes.

In the **new Federal Länder**, only two university institutes in Dresden (Saxony) took part in the Second Framework Programme (cf. Figure 3.5.1). The rigid procedures of the "call for tenders" in EC programmes - project proposals cannot be submitted continuously - prevented more institutions from the new Länder from taking part following German unification in 1990.

3.6 Summary and Conclusions

An analysis of the structure of participants - measured according to the number of project participations - clearly shows that in Germany it is primarily large companies, higher education institutes and non-university R&D establishments that have taken part in the Second Framework Programme. The **participation share of German SMEs** in EC projects (7.9%) is still low in comparison with the share of SMEs for the European Community as a whole (15.4%).

If the structure of participants is considered relative to the financial contributions of DG XII to individual project participants, there is a shift away from HEIs towards the group of BIGs, as these receive an above-average amount of EC funding per project participation. As only DG XII publishes its funding statistics, these statements apply only to EC projects administered by this Directorate General. Compared with the project funding of the Federal Ministry for Research and Technology (BMFT) related to special programmes, the share of EC funding going to SMEs is slightly higher (DG XII only).

German participants in the Framework Programme II obviously form a definite "community" which, once the circle of EC RTD support has been entered, take part repeatedly in various different project consortia. This makes it markedly more **difficult for first-time applicants** to gain an entry to this circle, and it is for SMEs that this constitutes the biggest barrier. Their participation is limited on average to approximately one EC project.

The fact that the participation rate of SMEs in the Second Framework Programme is still low is certainly due to a considerable extent to the **specific innovation patterns of SMEs** which, at least up to a point, run contrary to the orientation of the EC's Second Framework Programme towards precompetitive research and technological development. The R&D activities of SMEs are concentrated more on development and less on research, which is mainly carried out by large companies (cf. Meyer-Krahmer, 1989, p.24ff). Innovations in SMEs are determined to a high degree by the market, and because of their limited financing capability SMEs have to be very sure that their results will be usable. In its precompetitive RTD support (further from the market), directed towards technological innovations, the Framework Programme II has oriented itself towards the innovation patterns of BIGS, HEIs and RDIs; this orientation has overestimated the available R&D resources of many SMEs.

Besides the specific innovation pattern, however, there are **other reasons** for low participation of SMEs; on the one hand EC projects are often too large and too much directed towards scientific excellence for the orientation of SMEs. On the other hand, SMEs especially have **difficulties** in forming **R&D networks** across the EC or gaining access to such networks.

However, EC RTD support should not overlook the division of labour that exists in the macroeconomic innovation process between large companies and SMEs: **innovating SMEs play an important role** not only in the innovation phase but also in the diffusion of complex basic technologies (cf. Meyer-Krahmer, 1989, p.30). Should EC RTD policy wish not only to promote the generation of new technologies, but to speed up their diffusion, SMEs would represent an important target group; this requires the creation of SME-specific RTD programmes.

Regarding the various RTD programmes, **German participation is clearly concentrated in the five programmes** ESPRIT, BRITE/EURAM, SCIENCE, JOULE and RACE. The share of German participants in ESPRIT, BRITE/EURAM, SCIENCE and JOULE is above the average for Europe as a whole and is a particular characteristic of Germany. On the other hand, the participation of German R&D institutions in the programme lines "Biotechnology" and "Energy" is below average.

The **regional distribution of German participants** reflects the locations of research-intensive large companies, technical higher education institutes and the institutes of non-university R&D organisations. There are virtually no participants from the new Länder in the Second Framework Programme. Feedback from the written questionnaire addressed to non-participants shows, however, that R&D institutions in the new Länder have high expectations from the RTD support of the EC.

4. Importance of EC RTD Policy and its Influence on Enterprises

4.1 Characteristics of Participants and Non-participants in the Survey

209 of the enterprises participating in the Second Framework Programme completed our written questionnaire. However, they **did not always answer all questions**, so that in some questions there are slight deviations from the figure of 209 replies. Two participating and six non-participating enterprises did **not** answer the question on firm size categories; thus in the tables the disaggregation of responses into BIGs and SMEs relates to 207 participating and 271 non-participating firms (cf. Table 4.1.1). The sum of the answers of BIGs and SMEs when separately given is thus smaller than those of the whole group of enterprises together.

Table 4.1.1: Size of Firms investigated

Full time Employees	Participants		Non-participants	
	absolute	%	absolute	%
up to 99	61	29.2	130	46.9
100 - 249	14	6.7	55	19.9
250 - 499	14	6.7	32	11.6
500 - 999	12	5.7	18	6.5
1000 - 1999	21	10.0	27	9.7
2000 and more	85	40.7	9	3.2
no particulars	2	1.0	6	2.2
Total	209	100.0	277	100.0

Of this sample of enterprises (n = 207), 40.6% were small and medium-sized firms with less than 500 employees and 56.4% were BIGs with 500 and more employees (cf. Table 4.1.1). The firms that replied in the group of EC participants came mainly from the **sectors** of electrotechnics (28.2%), service companies (13.4%), mechanical engineering (11.5%) and the aerospace industry (11.0%).[25] Regarding **regional distribution** the sample of participating firms shows concentrations in the regions (NUTS 2 classification) of Oberbayern (16.3%), Darmstadt (11.0%), Köln (9.1%) and Stuttgart (8.6%).

25 Because of the relatively small number of answers and the wide range of industrial sectors involved it was not considered appropriate to analyse enterprise questionnaires according to branches.

Regions in the new Federal Länder are not represented in the sample of participating firms (cf. Table A4.1.1 in the Annex).

Of the ca. 1,200 firms written to as **non-participants**, there were 277 replies from firms who had not yet participated in EC projects. Only **three enterprises had already submitted a project application** to the EC, but had been turned down; all the other firms had not yet submitted any application for project support to the EC. The non-participating firms differed markedly from the participating sample regarding **number of employees**: 78.2% of non-participating firms belonged to the category of small and medium-sized enterprises with less than 500 employees, 19.4% were BIGs (cf. Table 4.1.1). The structure of the non-participating firms regarding firm size is not equivalent to the structure of the EC participants; rather, it is equivalent to the structure of innovating firms in the Federal Republic of Germany. Based on a study sample of 11,000 innovating enterprises, statistics of the SV-Wissenschaftsstatistik calculated a share of 90.2% for innovating firms in the size category up to 500 employees, and 9.8% for innovating BIGs in 1987 (cf. SV-Wissenschaftsstatistik 1990, p.13).

The **regional distribution** of non-participants is concentrated in Oberbayern, Stuttgart, Düsseldorf and Karlsruhe. Enterprises in regions of the new Länder were also included; these constituted approximately 18% of the sample (cf. Table A4.1.1 in the Annex). The **sectoral distribution** of non-participants showed concentration in the branches of mechanical engineering (31.8%), electrotechnics (20.9%), the chemical industry (8.7%), precision engineering/optics (5.8%) and service companies (4%).

Non-participating firms showed considerably less involvement in **publicly funded R&D programmes** than participating firms (40.8% as opposed to 82.3% - cf. Table 4.1.2). Obviously the non-participants are not sufficiently aware of the programmes, or do not fulfill the requirements for participation. If firms in the non-participant sample do take part in RTD programmes, these are by far most often programmes of the German Federal Ministry for Research and Technology (BMFT), followed by programmes of the Länder. For the group of EC project participants, BMFT programmes are by far the most important, with Länder programmes a long way behind. Programmes of the Federal Ministry for Economic Affairs (BMWi) and others are of secondary importance for both groups. In both groups, too, the BIGs participate markedly more than the SMEs in publicly funded R&D programmes.

EUREKA does not represent an alternative to EC research funding: with two exceptions, firms in the non-participant group are not taking part in EUREKA projects (cf. Table 4.1.2). EC participants, on the other hand, show a considerably higher degree of involvement in EUREKA projects (25%). This seems to indicate that EUREKA projects appeal to the same potential participants as EC RTD programmes.

Table 4.1.2: Is Your Organisation Participating in (other) Publicly Supported RTD Programmes?				
RTD-Programmes	Participants		Non-participants	
	absolute	%	absolute	%
No/no particulars	37	17.7	164	59.2
Yes, namely:	172	82.3	113	40.8
other EC projects	126	73.3	-	-
Eureka projects	43	25.0	2	1.8
BMFT support programmes	131	76.2	72	63.7
BMWi support progammes	28	16.3	22	19.5
Länder programmes	44	25.6	34	30.1
Other programmes	18	10.5	10	8.8

Multiple answers were possible.

When the samples of **non-participating enterprises and enterprises participating in EC projects** are compared for number of employees, regions, sectors and involvement in research support programmes, **substantial differences are found in all four indicators**. The structure of the group of non-participant firms does not reflect the structure of the participant group; rather, it complements it. The great majority of non-participants are SMEs; they come from the sectors of mechanical engineering, the chemical industry and precision engineering/optics, and take less part in publicly funded research programmes. By contrast, the group of participants reflects the structure of large companies in the branches of electrotechnics and the aerospace industry.

4.2 Description of Sample and Comparison with Basic Population

Altogether 826 enterprises from the Federal Republic of Germany took part in EC projects in the Second Framework Programme of the CEC, of which 209 answered our written questionnaire. These make up our sample of participating firms. The 826 enterprises represent a proportion of 39.2% of the basic population of all German participants in the EC programmes. In comparison with this, the group of enterprises in our sample of participants is slightly under-represented, with a share of 35.7%. The proportions of BIGs and SMEs in the sample also deviate from the basic population in the sense that there is a **higher proportion of SMEs in our survey sample than in the basic population** of participating enterprises (cf. Table 4.2.1). When interpreting the data from the sample of participating firms, therefore, it should be borne in mind that SMEs influence the results more strongly than would be the case for the basic population. The reason for this difference in the composition of the sample is to be found in the different frequency of involvement of BIGs in EC projects, and the **resultant differences in answering behaviour:** large companies take part more often in EC projects, and they did not answer the questionnaires for all of their EC projects.

Table 4.2.1: Comparison of the Basic Population of Participating Firms with the Survey Sample

	Basic Populations of Participating Firms (n= 826)		Survey Sample of EC Participants (n= 207)	
	absolute	%	absolute	%
Large firms	660	79.9	118	57.0
SMEs	166	20.1	89	43.0
Total	826	100.0	207	100.0

When a comparison is made of **positions in the consortium**, the proportion of **main contractors** is considerably smaller in the basic population than in the survey sample, both for BIGs (basic=25.0%, sample=41.9%) and SMEs (basic=20.7%, sample=36.0%). Contractors, on the other hand, are more strongly represented in the basic population than in the sample both for BIGs (basic=67.0%, sample=52.1%) and for SMEs (basic=61.4%, sample=56.2%). The reason for this distortion is to be found in the better answering behaviour and availability of addresses of main contractors. [26]

A **comparison** according to **participation in EC programmes** shows the following distortions (cf. Table 4.2.2):

- Participants in the **BRITE/EURAM** Programme are more strongly represented in the sample than in the basic population;

- **ESPRIT** and **RACE**, on the other hand, are under-represented in the sample.

26 On this aspect cf. also Chapter 1 (Methods) and the description of problems in compiling the basic population: due to insufficient quality of the addresses of EC participants, and the fact that main contractors were more identifiable, it was more often possible to write to them than to contractors or associates.

Table 4.2.2: Structure of the answering Firms in EC Programmes				
EC-Programmes (Second Framework Programmes 1987-1991)	**Survey Sample (n=209)**		**Basic Population (n=826)**	
	absolute	**in %**	**absolute**	**in %**
1. Quality of Life	**0**	**0.0%**	**0**	**0.0%**
e.g.: Human Gen. Analysis	0	0.0%	0	0.0%
Medicine/Health	0	0.0%	0	0.0%
Radiation Protection	0	0.0%	0	0.0%
STEP/EPOCH	0	0.0%	0	0.0%
2. Information and Communication Technologies	**90**	**43.1%**	**499**	**60.4%**
e.g.: AIM	2	1.0%	10	1.2%
DELTA	3	1.4%	9	1.1%
DRIVE	12	5.7%	49	5.9%
EURET	0	0.0%	0	0.0%
ESPRIT II	55	26.3%	324	39.2%
RACE I	18	8.6%	107	13.0%
3. Modernisation of Industry	**82**	**39.2%**	**215**	**26.0%**
e.g.: B.C.R	1	0.5%	0	0.0%
BRITE/EURAM	80	38.3%	203	24.6%
Raw Materials	1	0.5%	12	1.5%
4. Biological Resources	**4**	**1.9%**	**8**	**1.0%**
e.g.: Agriculture	0	0.0%	0	0.0%
BIO/BRIDGE	2	1.0%	3	0.4%
ECLAIR	2	1.0%	5	0.6%
FLAIR	0	0.0%	0	0.0%
5. Energy	**24**	**11.5%**	**96**	**11.6%**
e.g.: Decommissioning	7	3.3%	18	2.2%
Fusion	0	0.0%	0	0.0%
JOULE	14	6.7%	48	5.8%
Radioactive Waste	1	0.5%	8	1.0%
TELEMAN	2	1.0%	7	0.8%
6. Science and Technology for Development	**1**	**0.5%**	**0**	**0.0%**
7. Marine Resources	**1**	**0.5%**	**2**	**0.2%**
e.g.: FAR	0	0.0%	0	0.0%
MAST	1	0.5%	2	0.2%
8. Improvement of European S/T Cooperation	**1** °	**0.5%**	**6**	**0.7%**
e.g.: DOSES	0	0.0%	0	0.0%
EUROTRA II	0	0.0%	1	0.1%
Large-scale facilities	0	0.0%	0	0.0%
MONITOR	0	0.0%	0	0.0%
SCIENCE	1	0.5%	0	0.0%
SPES	0	0.0%	0	0.0%
VALUE	0	0.0%	5	0.6%
9. Other (related) Programmes	**6**	**2.9%**	**0**	**0.0%**
Total	**209**	**100.0%**	**826**	**100.0%**

Thus when applying the results of the survey study sample to the population of participating enterprises as a whole, it must be borne in mind that SMEs and main contractors are more strongly represented in the sample than in the basic population; moreover the share of BRITE/EURAM participants is larger, and the share of ESPRIT and RACE participants smaller, in the sample than in the basic population.

4.3 Importance and Strategic Significance for Participating Enterprises

4.3.1 Characterisation of EC Projects

Large enterprises (BIGs) mainly carry out their EC projects in the R&D department of one of their groups, in the R&D department of the enterprise, or in the central R&D unit of the company. SMEs usually conduct their projects in their production unit or cross-departmentally in a project team, as they often do not have separate R&D departments.

Both for BIGs and for SMEs, EC projects are **characterised primarily as applied research** (52.5% and 41.6% respectively) and **secondly as experimental development** (27.1% and 25.8% respectively). In addition, the EC projects of SMEs are characterised by demonstration projects and diffusion of new technologies (31.5%). Basic research does not play a role in the EC projects of enterprises (cf. Table 4.3.1.1). There are only slight differences between the typification of the projects as a whole and the typification of the parts of the projects undertaken by the enterprises.

Table 4.3.1.1: Which Type characterises Best Your Share in the Whole Project ?

	All Firms		Large Firms		SMEs	
	absolute	%	absolute	%	absolute	%
Basic research	4	1.9	3	2.5	1	1.1
Applied research	101	48.3	62	52.5	37	41.6
Experimental development	55	26.3	32	27.1	23	25.8
Demonstration and diffusion	49	23.4	21	17.8	28	31.5
Total	209	100.0	118	100.0	89	100.0

Regarding time horizons, both BIGs (64.4%) and SMEs (59.6%) are oriented towards **medium-term economic exploitation of the project results within three years** after completion of the project. Approximately one quarter of the BIGs have a long term project orientation, whereas about one quarter of SMEs are aiming at short term economic exploitation of project results (cf. Table 4.3.2.1).

The average **number of partners** in EC project consortia is approximately 6.6; the average for SMEs is 5.3 and for BIGs 7.7 partners (cf. Table 4.3.1.2). **Enterprises in EC projects cooperate** most frequently **with other enterprises**, and especially with enterprises from other EC countries. Higher education institutes and non-university R&D institutes were less in demand as partners from enterprises.

Unfortunately, no statements can be made for participating firms on the financing and scientific capacities of the projects, as the answers received were too varied and did not observe the size categories.

Table 4.3.1.2: How many Partners and Associates have participated in this EC Project?

	Large Firms		SMEs	
	M	se	M	se
Firms from Germany	1.3	1.7	1.0	1.3
Firms from EC countries	3.4	4.3	2.1	2.0
Universities from Germany	0.4	1.1	0.4	0.6
Universities from EC countries	1.0	1.4	0.7	0.9
RD Institutes from Germany	0.3	0.6	0.3	0.6
RD Institutes from EC countries	0.8	1.2	0.6	0.9
Other organisations from Germany	0.1	0.5	0.1	0.2
Other organisations from EC countries	0.1	0.7	0.1	0.4
Total	7.7	6.4	5.3	3.8

M = Mean; se = standard error

4.3.2 Strategic Role of EC Projects

The strategic importance of EC projects for participating enterprises can be characterised as follows (cf. Table 4.3.2.1):

1) **R&D project portfolio**: for about two-thirds of BIGs (64.7%) and SMEs (60.2%), the EC project is an **integrated component of their R&D project portfolio**; thus it follows that the EC projects are interrelated with other in-house R&D projects. Only one quarter of replies describe connections with other research projects as slight; only in a few cases is there no connection with other projects.

2) **Technological competence**: regarding technology competence, EC projects of **SMEs** tend to be in their present **strategic core area** (47.7%); 17% of SMEs are aiming at a new potential area of competence. For **BIGs** the result is **very ambivalent**: from the viewpoint of these enterprises the EC project tends to be placed in "**one of several**" technology areas (37.9% of BIGs) and not so often in the present strategic core area (28.4%); however one quarter of BIGs (25.9%) are aiming at a potential new area of technology in their EC project participation, and only for a few (7.8%) is the project of marginal technological importance.

3) **Company strategy**: in the strategy of BIGs (54.2%) and of SMEs (64.7%), the EC project serves the purpose of **expansion of an existing business field**; however, only one fifth of SMEs (21.2%) and one sixth of BIGs (15.9%) are aiming to establish a new field of business with the help of the EC project. Reduction of costs (14%) and diversification (11.2%) play some part for BIGs only.

The **ambivalent significance of EC projects for the technology competence of BIGs** is reflected in interviews with the heads of R&D departments and EC project leaders: some experts regard the technological level of some EC projects as "average", and below the level of their own R&D; according to them, the gain in technological competence is small. These statements refer to the project as a whole. For instance in one EC project (RACE), one project leader explained that "...with the aim of developing a demonstration plant in full view of the partners, continuous fears of a "know-how drain" arose; for this reason **only already-published methods and technologies** were brought into the project and no new elements were elaborated. Moreover, during the

Table 4.3.2.1: Composition of the R&D Portfolio (Firms)

	All Firms		Large Firms		SMEs	
	abs.	%	abs.	%	abs.	%
What Connection does the EC Project have to other R&D Projects in your Organisation?						
Integrated part of portfolio	129	62.6	75	64.7	53	60.2
Few connections to other R&D projects	55	26.7	29	25.0	26	29.5
No connections to other R&D projects	22	10.7	12	10.3	9	10.2
Total	206	100.0	116	100.0	88	100.0
Time span of this Project						
Short-term	34	16.3	13	11.0	21	23.6
Medium-term	131	62.7	76	64.4	53	59.6
Long-term	44	21.1	29	24.6	15	16.9
Total	209	100.0	118	100.0	89	100.0
What Role does the EC Project play for your Technological Competence (Technical Standing)?						
Strategical core area	75	36.4	33	28.4	42	47.7
Potential new core area	46	22.3	30	25.9	15	17.0
Just one technology area among others	67	32.5	44	37.9	23	26.1
Rather peripheral technology area	18	8.7	9	7.8	8	9.1
Total	206	100.0	116	100.0	88	100.0
What Role does the EC Project play in your Company Strategy?						
Establish new business areas	36	18.6	17	15.9	18	21.2
Extension of a business area	113	58.2	58	54.2	55	64.7
No significance	8	4.1	5	4.7	3	3.5
Diversification	18	8.6	12	11.2	6	7.1
Cost reduction	19	9.8	15	14	3	3.5
Total	194	100.0	107	100.0	85	100.0

running time of the project the technological goals were overtaken by the general development of technology." In another example, the EC project (ESPRIT) was in the strategic core area of the company and, according to the statement of the head of the R&D department or project leader, served "...to speed up research in this area and build up the technological leading-edge of the enterprise; at the beginning of the project many of the technological goals appeared too speculative, causing reservations on the part of those responsible in the EC; it was possible to remove these reservations, and the fear of mutual knowledge drains, by structuring the project into subordinate aims and strictly separated task modules and results modules."

The financing of R&D projects by the EC is very important for enterprises: in 44.3% of SMEs and 34,5% of BIGs the project would not have been carried out without the EC support (cf. Table 4.3.2.2). The EC funding also contributed to the **projects** being carried out **more quickly and in extended form**. Financial support by the EC obviously plays an important role, especially for the SMEs. Only in **exceptional cases** would firms have been prepared to carry out the project **without EC support**: less than one tenth of BIGs (7.8%) and SMEs (8.0%) would have tackled the R&D project even without funding from the CEC.

Table 4.3.2.2: What Role did the EC Funding Play ?

	All Firms		Large Firms		SMEs	
	abs.	%	abs.	%	abs.	%
Independent of funding	17	8.3	9	7.8	7	8.0
Faster through funding	63	30.6	37	31.9	26	29.5
Larger scale possible through funding	85	41.3	50	43.1	34	38.6
Other goals without funding	25	12.1	16	13.8	8	9.1
Not at all without funding	80	38.8	40	34.5	39	44.3
Total	206	100.0	116	100.0	88	100.0

Multiple answers were possible.

The importance of the EC funding for the **annual R&D budget of the enterprises** varies: for the enterprises as a whole, EC funding constitutes an average proportion of 12.6% of the R&D budget. For **BIGs** the average proportion of EC funding is 6.6% (cf.

Table 4.3.2.3) and varies between 1% and 7.5% measured by the interquartile range.[27] For SMEs, EC support constitutes an average proportion of 20.1% of their R&D budget, and varies between 5% and 30% measured by the interquartile range.

In order to avoid misinterpretations it must be borne in mind that these figures were certainly **approximate estimates** by those answering the questionnaires and - as emerged from the interviews with firms - tended to be **overestimates**. Moreover, information from the BIGs does not relate to the company as a whole, but only to the division carrying out the project, or to the central R&D laboratory. According to our interview partners from BIGs, the EC funding constitutes a "drop in the ocean" relative to the total R&D budget of companies and can be placed, according to their statements, at around 1 percent. However, the R&D units carrying out the projects may view the situation quite differently:[28] the central R&D laboratory of an enterprise in the construction sector with ca. 30 employees was able to have 3-4 employees working on an EC project for a period of 4 years. The R&D budgets of **SMEs**, on the other hand, may be more strongly dependent on EC funding during the running period of the project: one small firm with a total workforce of 8 (4 of them in R&D) was able to finance 1.5 employees for the R&D project for its duration (36 months).

More than half the firms answering the questionnaire (55.3%) estimate that **in-house funding** for the firm's R&D projects will remain about the same in future (cf. Table 4.3.2.3). One third (31.8%) expects **EC RTD funding** to go up, whereas one quarter (26.9%) expects an increase in **German federal funding**.

---

27 The quartile interval is a statistical measurement of frequency distribution which is not so dependent upon extreme values and is thus appropriate for this question. It gives the size of the area between the first quartile (25% of answers) and the third quartile (75% of answers): for large enterprises, 50% percent of characteristics (share of EC funding in R&D budget) are between 1% and 7.5%, and for SMEs between 5% and 30%; in this way the effects of extreme values, which would disproportionately raise or lower the mean percentage, are excluded.

28 Jäger (Siemens) estimates that the European information technology firms received ca. 10% of their R&D expenses for long-term research and development in this field in 1990 from the ESPRIT II Programme. The contribution of ESPRIT II was significant for these enterprises in the information technology field, but made up only a small share of the total R&D effort of large enterprises; the responsibility for innovations therefore lies unmistakably with industry (Jäger 1992, p. 58).

Table 4.3.2.3: What Share in the Financing of your R&D Budgets do the following have ?

	Mean 1991 in %		in future decreasing		in future constant		in future increasing		Total	
	BIG	SME	abs.	%	abs.	%	abs.	%	abs.	%
Own funds	58.9	49.2	33	21.7	84	55.3	35	23.0	152	100
Own contracts	14.6	5.3	12	16.2	37	50.0	25	33.8	74	100
Contracts from firms	4.5	11.4	8	11.0	27	37.0	38	52.1	73	100
Contracts from HEIs/RDIs	0.1	1.6	6	17.1	27	77.1	2	5.7	35	100
EC Funding	6.6	20.1	39	26.4	62	41.9	47	31.8	148	100
Federal Republic funding	13.1	7.0	37	31.1	50	42.0	32	26.9	119	100
Länder funding	1.2	3.3	12	19.0	32	50.8	19	30.2	63	100
Other sources	1.0	2.0	9	17.3	37	71.2	6	11.5	52	100
Total	100.0	100.0								

4.4 Reasons and Aims of Participation, Results and Success Factors

Reasons for EC Participation

The German firms questioned are carrying out projects supported by the EC for two main reasons (cf. Table 4.4.1): firstly, they consider it very important to engage in a **European cooperation in their field of technology** (60.7% of BIGs and 61.4% of SMEs). Secondly, participation in the EC actions represent an **additional source of financing for** the firm's **research projects** (57.3% of BIGs and 61.4% of SMEs). For BIGs (57.3%) it is also important that the **subjects** of EC programmes **complement existing German support programmes**. Nearly half of the SMEs consider that support by the European Community has made possible a **longer-term orientation** of their R&D projects.

Table 4.4.1: Why are you carrying out R&D Projects with EC Funding ?

	All Firms		Large Firms		SMEs	
	abs.	%	abs.	%	abs.	%
Further financing source	121	58.5	67	57.3	53	60.2
Enables long-term orientation	103	49.8	61	52.1	42	47.7
Necessary supplement to German support	108	52.2	67	57.3	40	45.5
European cooperation is important	127	61.4	71	60.7	54	61.4
German programmes unsuitable	22	10.6	13	11.1	9	10.2
Exploration of new business areas	55	26.6	31	26.5	23	26.1
Establish European collaborations	93	44.9	52	44.4	40	45.5
Other reasons	13	6.3	9	7.7	3	3.4
Total	207	100.0	117	100.0	88	100.0

Multiple answers were possible.

Project Aims and Results according to Firm Size

The primary aim of the concrete EC projects is to **increase the knowledge base**. This is equally true for large companies (76.3%) and SMEs (67.8%). The second most important aim is **better connections with European enterprises** (55.3% of BIGs and 46% of SMEs); cooperation with European higher education institutes and R&D institutes (33.3% of BIGs and 34.5% of SMEs) plays a markedly smaller role. **SMEs** (58.6%) tend to be pursuing **innovative product development** in their EC projects more often than BIGs (41.2%), whereas large companies are aiming more at **product improvements** (50.9% as opposed to 43.7% of SMEs). Another important goal for BIGs (40.4%) and SMEs (42.5%) is **to increase the qualification of R&D personnel**, whereas increasing the number of personnel and enhancing scientific reputation are not regarded as important. The **acquisition of generally applicable methods and instruments** is important for large companies (40.4%), for SMEs (42.5%) the **design of a prototype** constitutes another important project aim. Innovations in processing technology and in the area of services are not high-ranking goals in EC projects, indicating that the projects are more generally oriented towards product innovations.

88.0% of firms state that they have **attained their initial technological and/or scientific aims**. Approximately three quarters of the large companies (75.7%) and SMEs (71%) questioned have achieved 80% to 100% of their project aims. This clearly contradicts the assertions often made in public that EC projects are very inefficient and do not produce many results.

Projects Aims and Results according to Selected Programme Lines

The aims pursued and the results obtained were examined and evaluated in more detail according to the various programme lines; for this analysis, EC programmes were combined **into three groups**:

- In the programme line "**Information and Communication Technologies (I&C Technologies)**" (especially ESPRIT, but also RACE and DRIVE), 43% of firms replying to the questionnaire are represented; the EC programme affiliation corresponds to the Second Programme Line of the Second Framework Programme.

- In the programme line "**Modernisation of Industry**" (with clear concentration in BRITE/EURAM), 39% of firms replying to the questionnaire are represented; the EC programme affiliation corresponds to the Third Programme Line of the Second Framework Programme.

- The remaining replies constitute less than 18% of all replies and are combined for statistical reasons into one group ("**other areas**"); replies concentrate mostly on the "Energy" programme line.

In projects in the programme line "**I&C Technologies**", firms are aiming at increasing their knowledge base, improving their contacts to European firms, innovative development of their products and qualification of their R&D personnel (cf. Figure 4.4.1). By contrast, in the programme line "**Modernisation of Industry**" the improvement of products ranks second, followed by cooperation with firms in the European Community. In the "**other areas**" group, product improvement is also the second most important aim, followed by cooperation with European firms and further participation in EC projects. **Increasing the knowledge base** is generally regarded as the most important aim, followed by **cooperation with enterprises** in the EC.

Fig. 4.4.1 Which of the following Goals did you aim for in this Project? (Enterprises)

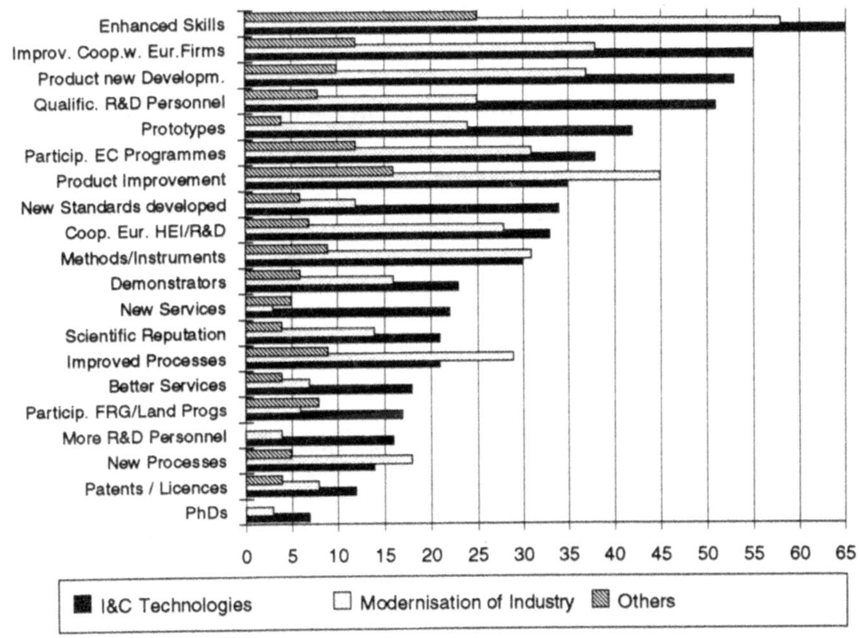

Examining the primary aims and their achievement in more detail, the aims of "**enhanced skills/increasing the knowledge base**", "**better cooperation with enterprises in the EC**" and "**higher qualification of R&D personnel**" have been directly attained in both programme lines and in the "other areas" group (cf. Figures 4.4.2 to 4.4.4). It is interesting in this context to consider the **time horizon**: whereas the aims listed above had already been attained at the time of completion of the project, this is not the case for the time horizon for product innovation. The **innovative development or improvement of products** is generally first achieved within the three years following project completion; in a few cases it has not yet been achieved. Particularly in the programme line "I&C Technologies", it is apparent that over two-thirds of firms replying to the questionnaire only expect product innovation to take place after a fairly long period of time has elapsed (cf. Figure 4.4.2).

The fact that not all aims can be realised directly on completion of the project also applies to **improved processing technology, design of prototypes, development of standards and new methods or instruments** (cf. Figures 4.4.2 to 4.4.4). These bar charts also emphasise the difference between setting up a demonstrator and designing a prototype: whereas the demonstrator plant is intended to demonstrate the technical feasibility of the R&D project, the design of a prototype already presupposes the intentions of manufacture and market introduction.

Thus it is necessary to take the **time factor** into account when making statements about the **impact of EC projects on improvement of the competitiveness of European industry**. As EC projects, for the enterprises involved, take on the character of application-oriented research, innovative or improved products cannot be expected straight away on completion of the project, and the possible impacts cannot be assessed at that point.

Fig. 4.4.2: Which Results did you achieve in this Project? (I&C-Technologies, Enterprises)

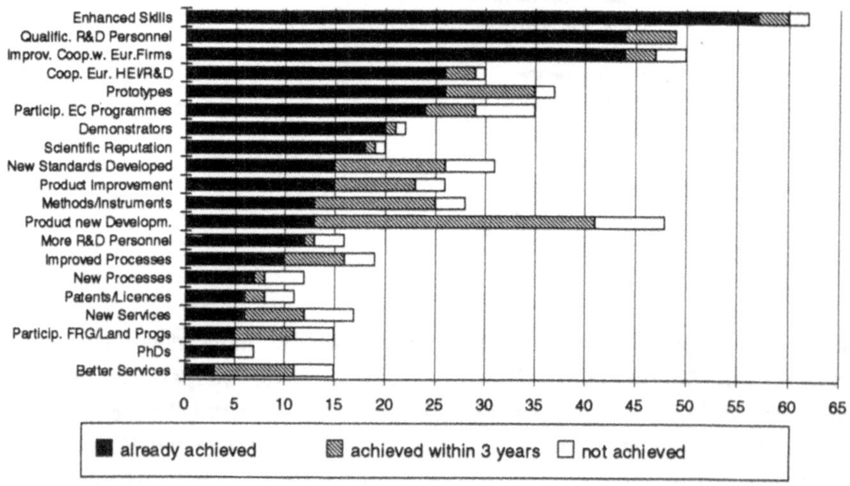

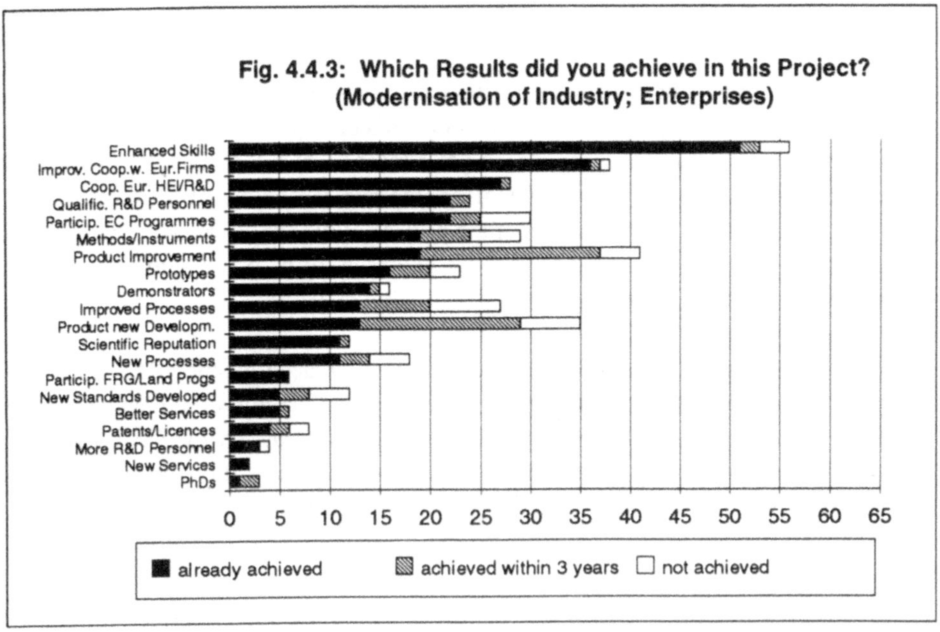
Fig. 4.4.3: Which Results did you achieve in this Project?
(Modernisation of Industry; Enterprises)
Enhanced Skills
Improv. Coop.w. Eur.Firms
Coop. Eur. HEI/R&D
Qualific. R&D Personnel
Particip. EC Programmes
Methods/Instruments
Product Improvement
Prototypes
Demonstrators
Improved Processes
Product new Developm.
Scientific Reputation
New Processes
Particip. FRG/Land Progs
New Standards Developed
Better Services
Patents/Licences
More R&D Personnel
New Services
PhDs
0 5 10 15 20 25 30 35 40 45 50 55 60 65
already achieved
achieved within 3 years
not achieved

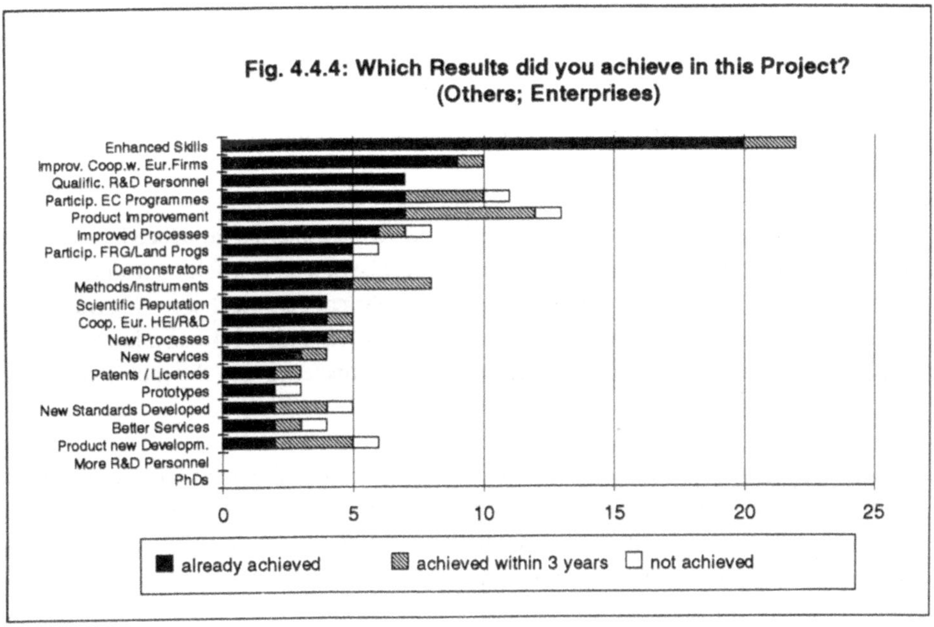
Fig. 4.4.4: Which Results did you achieve in this Project?
(Others; Enterprises)
Enhanced Skills
Improv. Coop.w. Eur.Firms
Qualific. R&D Personnel
Particip. EC Programmes
Product Improvement
Improved Processes
Particip. FRG/Land Progs
Demonstrators
Methods/Instruments
Scientific Reputation
Coop. Eur. HEI/R&D
New Processes
New Services
Patents / Licences
Prototypes
New Standards Developed
Better Services
Product new Developm.
More R&D Personnel
PhDs
0 5 10 15 20 25
already achieved
achieved within 3 years
not achieved

Factors Inhibiting and Favouring Success

What factors impeded or contributed to the success of the EC projects described in the questionnaires? One very important success factor is clearly the **adequate determining of technological goals** in the project: the goals must be such that they are in fact technologically attainable; the establishing of appropriate technological goals thus had a (very) strong influence on the success of the project. However, if partners changed their goals - and especially if they did so without reference to the other partners - this was (very) negative for the success of the project (cf. Figure 4.4.5).

Project management constituted another complex of determinants. The firm's own project management and the management by the main contractor were obviously (very) positive for the success of the projects. Project monitoring by the EC Commission, on the other hand, had only a limited importance for project realisation. However, it was named in the few instances of premature project termination as one of several contributory causes.

Other important determinants are grouped around the **cooperation** issue. The composition of the consortium and the technological competence of the partners were (very) positive for the successful realisation of the project (cf. Figure 4.4.5). Coordination between the partners regarding project content also favoured project success. Important obstacles were the expenditure of time and money on communication - associated with a large number of partners - , different attitudes regarding adherence to time schedules, and possible internal re-organisation of partners. The different cultures and mentalities of the partners, often referred to in public discussion as a problem in EC projects, had only a slight positive or negative influence on the success of projects according to the results of the survey.

A small number of interviewees indicated that the EC had exerted influence in the choice of partners. This did not apply to the project organisation as a whole, but only to individual partners. These consortia assembled for "**political reasons**" were regarded with scepticism, particularly by the main contractors interviewed, who considered that the "political" partners often lacked the technological competence and the management capabilities necessary for participation in the project.

Another important success factor emerged clearly from the interviews with project leaders and heads of R&D departments, regarding the structuring of work within the project consortium. The partners **structure or divide the tasks in the project** in such a way that the part of the project carried out by a firm can lead to an autonomous result for

Fig. 4.4.5: Which Factors have helped or hampered the Success of this EC Project or led to scrapping of Project?

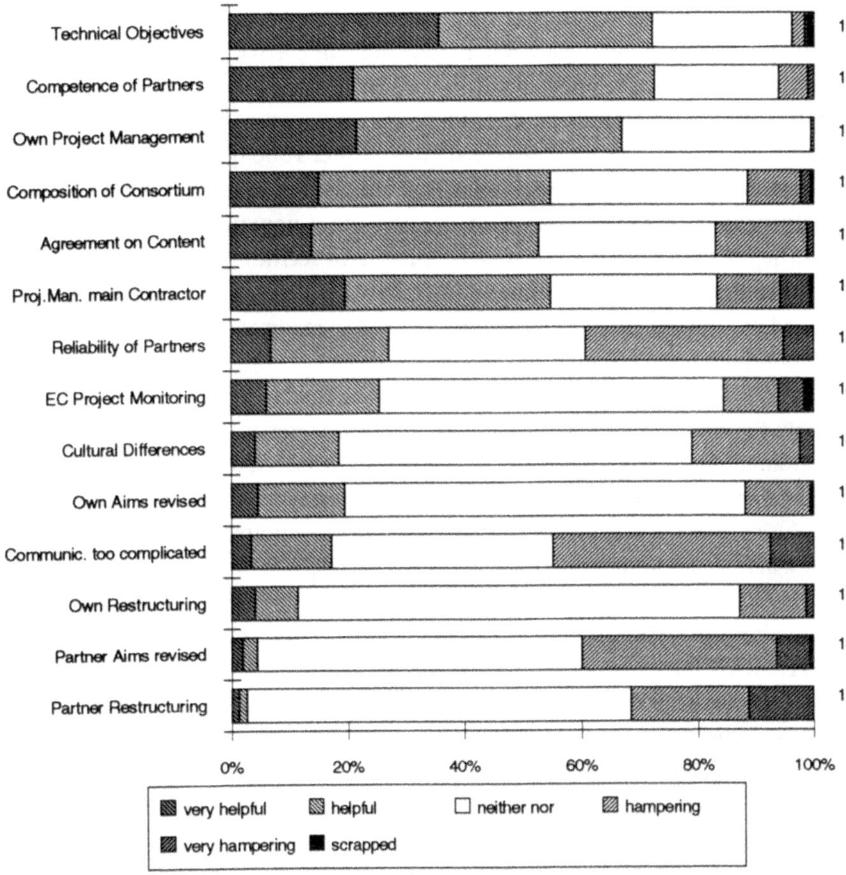

that firm. For instance, the result of the project may be a new product consisting of several components, each of which represents a product for one of the participating firms. With this **"modular"** approach each partner can aim at an autonomous result, and is less dependent upon the general quality of work within the consortium.

4.5 Impacts on Participating Enterprises

4.5.1 Economic, Strategic, Technical and Scientific Impacts

When the possible impacts of an R&D project are categorised as economic, technological, scientific and strategic effects, it can be established for the EC projects of German participants that - at any rate from the subjective viewpoint of those questioned - it was the latter two types of impact that primarily occurred. Among the four most important effects (regardless of whether these were described as very strong or less strong) were ranked **increased scientific capability** (as the most important scientific effect), the **establishing of new research fields** and a **change in technology strategy** (as the most important strategic effects) (cf. Figures 4.5.1.1 to 4.5.1.3). **Improved product quality** was named by firms as the most important (indirect) economic effect.

The **impacts show no substantial variations dependent on the EC programme lines**; only the order of the individual effects named varies to a considerable extent. In the programme line "**Modernisation of Industry**", which includes BRITE/EURAM, a programme strongly oriented towards new manufacturing technologies, the reduction of manufacturing costs is ranked sixth, as opposed to the programme line "I&C Technology", in which it is ranked ninth (cf. Figures 4.5.1.1 and 4.5.1.2).

The **following example** illustrates the impacts and the problems controlling them: one **BRITE/EURAM project** was concerned developing a new aluminium alloy and an optimised process technology for its production. From the strategic viewpoint, the purpose of the project was to build up a research area important to the firm, the area of "new materials", and to acquire technological competence. The technical, scientific and strategic aims were indeed achieved in the EC project. Firstly, the research work did produce an increase in technological competence. Secondly, the technology strategy in the area of "new materials" was changed, and new research fields were opened up: impulses from the EC project work caused the enterprise to reflect whether conventional materials might not in fact possess properties similar to those of the new materials. According to statements by the project leader, the market introduction of new materials is often very difficult, as the price levels necessitated by changes in processing technology are often not accepted.

Direct economic impacts such as **absolute increases in turnover** and a **higher market share in the EC and/or USA/Japan**, on the other hand, are mentioned by ca. 30% or less of firms replying to the questionnaire. If a higher market share has been achieved, this has occurred primarily in **markets within the EC** and less in the markets of Japan or the USA (cf. Figures 4.5.1.1 to 4.5.1.3). However, an assessment of the economic impacts of EC projects must take the time factor into account: almost two-thirds of firms replying to the questionnaire state that there have not yet been economic effects, but that they expect an increase in turnover or a higher market share **within the next three years or even later**. Since a certain time interval is necessary for the conversion of project results into new products (cf. Chapter 4.4), economic impacts cannot yet be empirically established for the Second Framework Programme.

Asked for their subjective opinion about whether the EC project had changed the technological position of the participating firm relative to competitors, about two-thirds of firms stated that their **technology position within Germany** (47.3%) **and in the EC** (60.3%) had improved. By contrast, less than half the firms questioned see an improvement in their technological position relative to the USA (50.0%) or Japan (43.8%). Clearly their participation in EC research projects does **not so much directly improve the technology position** of participating firms **relative to Japan and the USA**, but rather improves their position within the European Community and within Germany. Whether this improvement in technology position, however, has produced indirect positive competitive impacts for German firms by limiting imports could not be analysed.

Results of the questionnaire addressed to German firms show that participation in EC projects increases scientific capability and influences technology strategies. By **stimulation of the scientific and technological basis** - at least for a proportion of participating enterprises - **an important aim of the Second Framework Programme has been achieved**. The link targeted in the Second Framework Programme between strengthening innovative capability by means of RTD support, and improving the international competitiveness of the European Community in the USA/Japan triad, cannot yet be established in the empirical results of this study.

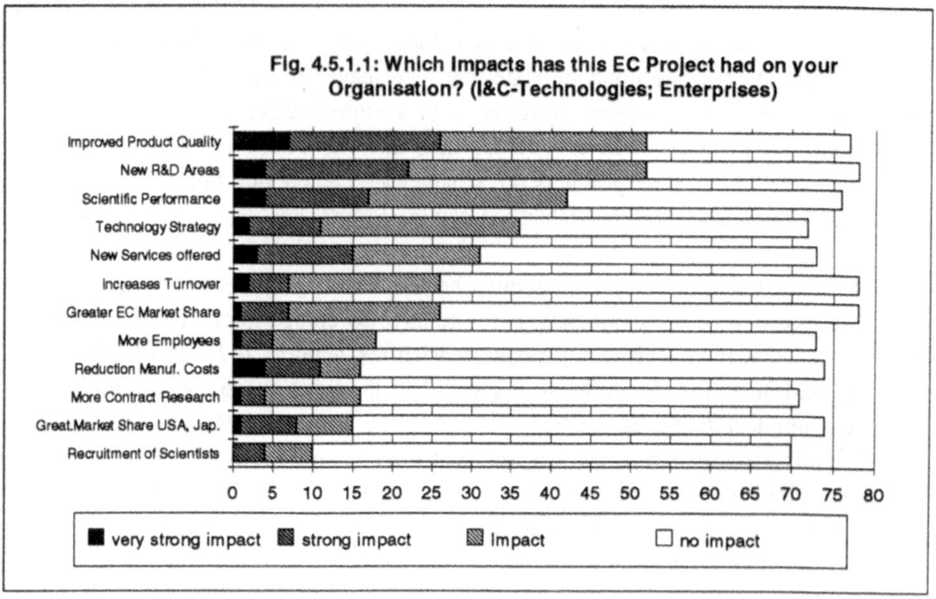

Fig. 4.5.1.1: Which Impacts has this EC Project had on your Organisation? (I&C-Technologies; Enterprises)
Improved Product Quality
New R&D Areas
Scientific Performance
Technology Strategy
New Services offered
Increases Turnover
Greater EC Market Share
More Employees
Reduction Manuf. Costs
More Contract Research
Great.Market Share USA, Jap.
Recruitment of Scientists
0 5 10 15 20 25 30 35 40 45 50 55 60 65 70 75 80
very strong impact
strong impact
Impact
no impact

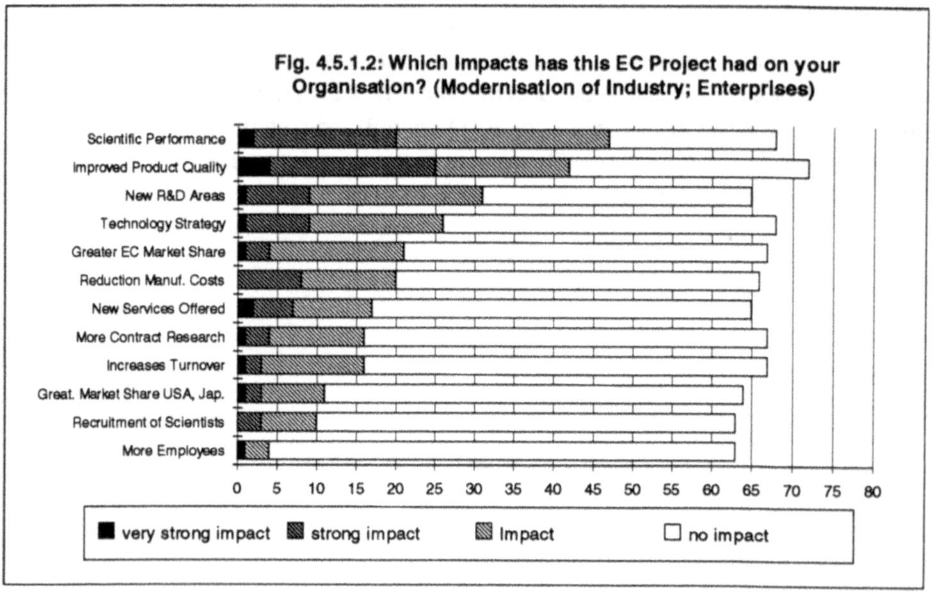

Fig. 4.5.1.2: Which Impacts has this EC Project had on your Organisation? (Modernisation of Industry; Enterprises)
Scientific Performance
Improved Product Quality
New R&D Areas
Technology Strategy
Greater EC Market Share
Reduction Manuf. Costs
New Services Offered
More Contract Research
Increases Turnover
Great. Market Share USA, Jap.
Recruitment of Scientists
More Employees
0 5 10 15 20 25 30 35 40 45 50 55 60 65 70 75 80
very strong impact
strong impact
Impact
no impact

Fig. 4.5.1.3: Which Impacts has this EC Project had on your Organisation? (Others; Enterprises)

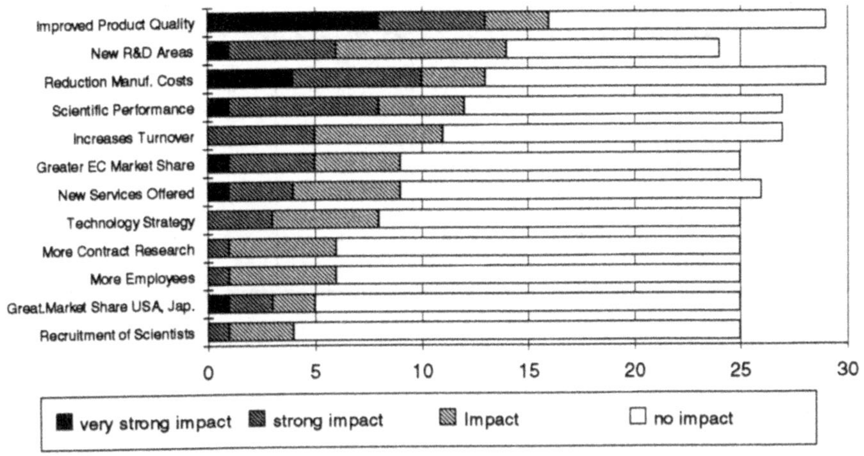

4.5.2 Impacts on the Cooperation Behaviour

13% of the firms questioned had never before been engaged in a cooperation in the area of R&D. Previous to their EC project participation, the majority of enterprises had already had **numerous cooperative R&D relationships** (87.0%). Previous cooperations had most often been with **other enterprises** such as suppliers, customers or competitors, with suppliers ranking first and customers second as critical cooperation partners. Other important partners were **higher education institutes** and **non-university R&D institutes** (cf. Table 4.5.2.1). **Business consultants** and **engineering bureaux** tend to play a subordinate role in cooperation in the area of R&D.

Regionally speaking, participating German enterprises cooperate most frequently with other enterprises (suppliers, n=81/customers, n=77/competitors, n=54) **in Germany**, but also with enterprises **in other EC countries**. Cooperative relationships with firms **in Europe outside the EC, in the USA and in Japan are limited**. The same regional ranking applies to cooperation with HEIs and with RDIs.

Table 4.5.2.1: Cooperation with other Organisations in R&D (Number of Participants)

	Germany	EC	Rest of Europe	USA	Japan	Rest of World
Suppliers	81	49	30	22	5	15
Customers	77	61	28	27	17	11
Competitors	54	67	9	16	10	5
Universities	144	66	20	14	5	4
RDIs	91	51	13	9	2	3
Advisors	57	19	3	6	1	2
Network	25	29	3	3	2	2

Multiple answers were possible.

Cooperation with European partners in the EC projects has increased the general **willingness to cooperate** of BIGs (74.6%) and SMEs (68.7%) (cf. Table 4.5.2.2). Because of their positive experiences in EC projects, more than half the enterprises have expanded their cooperation in the area of R&D, extending it especially to include cooperation in further EC projects. For four large companies, positive long term experiences have led to the setting-up of joint R&D establishments. One fifth of firms (20.2% of BIGs and 26.5% of SMEs) have expanded **cooperation into other business areas**, particularly product development. However, a favourable impact on cooperation in the areas of production, marketing and sales could not be established.

In 13% of the firms questioned, either general willingness to cooperate has been reduced (12.9%) due to **negative experiences**, or no further cooperations in EC projects are planned ((8.1%), as the cost/benefit relationship is not clearly favourable (cf. Table 4.5.2.2).

All in all, EC research support can definitely be said to have increased the readiness of firms to cooperate. In general this is mainly limited, however, to cooperation in further projects within RTD support by the European Community; the extension of cooperation to other areas is less frequent.

Since many firms have already engaged previously in cooperations, EC RTD support programmes do not so much initiate totally new cooperative relations as support the **stabilising and reinforcement of cooperative relationships** at the level of the European Community. Personal interviews with representatives of firms confirmed this fa-

Table 4.5.2.2: What Impacts did your Cooperation with European Partners have?

	Large Firms		SMEs		All Firms	
	abs.	%	abs.	%	abs.	%
No further cooperations	7	5.9	10	11.2	17	8.1
Decreased willingness to cooperate	19	16.1	7	7.9	27	12.9
Increased willingness to cooperate	85	72.0	57	64.0	143	68.4
Extension of R&D cooperation	60	50.8	44	49.4	105	50.2
Extension of cooperation	23	19.5	22	24.7	45	21.5
Total	118	100.0	89	100.0	209	100.0

Multiple answere were possible.

vourable, stabilising element of European cooperation: many of the partners are known to one another from previous contacts, and apply to the EC's call for proposals as a ready-formed consortium or as the "core" of a consortium in which only a few of the partners are "new". In the opinion of interviewees, EC RTD support gives **SMEs** an especial opportunity to expand existing cooperations into areas other than R&D (e.g. joint sales or service), or to extend business relationships between suppliers and customers into the area of applied research and development.

Existing contacts and **cooperation in other EC projects** are important for firms when **selecting partners** (cf. Figure 4.5.2.1). Conferences and publications by partners have a limited importance. The EC information data banks and the German federal and Länder public liaison offices play only a marginal role in the search for partners. According to statements in interviews with experts, one main reason for this is that although data banks and liaison offices can provide sufficient data quantitatively speaking, this does not suffice for a qualitative assessment of potential partners, nor for the "forging of networks". Personal contacts and cooperation experiences cannot be measured in terms of quantities of data; they still remain the best way of finding partners.

These statements apply generally to large enterprises and to SMEs. However, if results are analysed more intensively and in more detail, it can be ascertained that **personal contacts or personal mediation by other partners is even more important for SMEs**. BIGs, on the other hand, make better use of publicly available sources of information, and place more emphasis on the importance of EC information data banks, public liaison offices and the EC offices of scientific organisations.

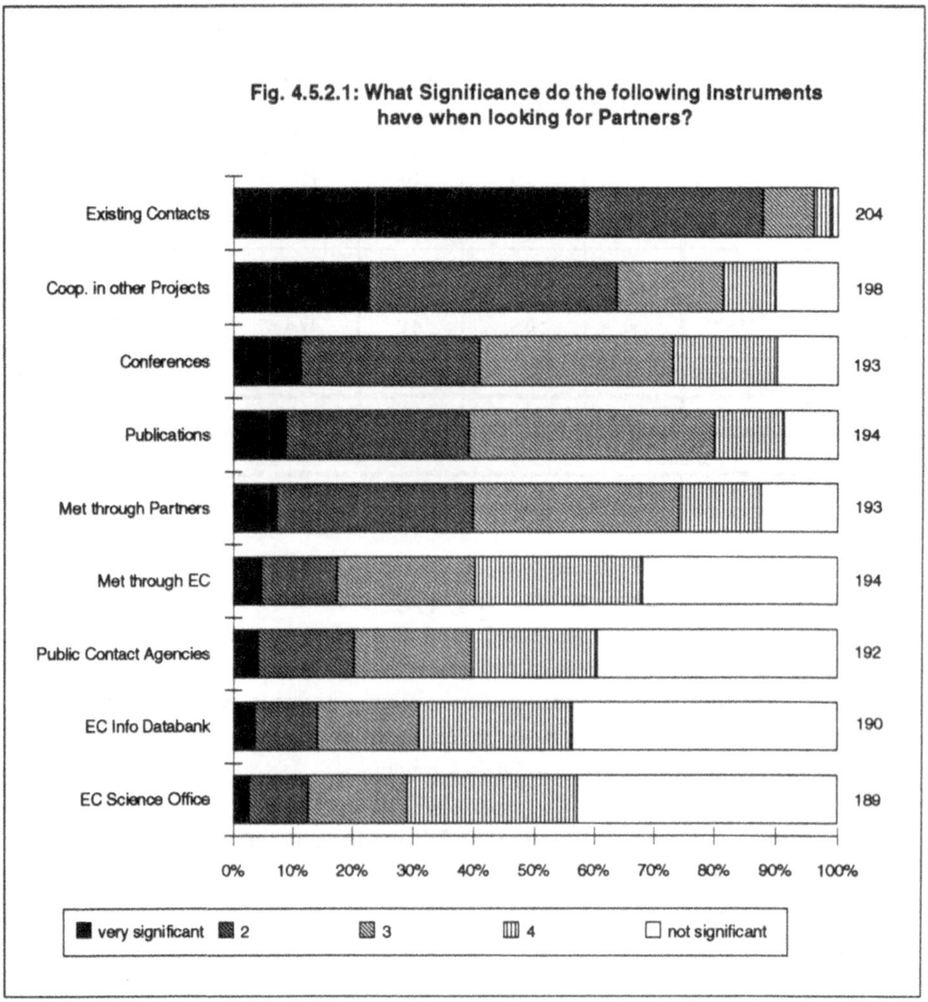

4.5.3 Social and Environmentally Relevant Impacts

Apart from their economic, scientific, technical and strategic impacts, an average of more than two-thirds of the firms questioned see **no social or environmentally relevant impacts** of the EC project. At most, there are small improvements in working conditions, in environmental protection and in energy saving (cf. Figure 4.5.3.1). The EC projects obviously lead to improvements in these areas only if the project itself has explicit, social or environmentally relevant aims.

Fig. 4.5.3.1: Has this Project had any further social or environmentally relevant Impacts?

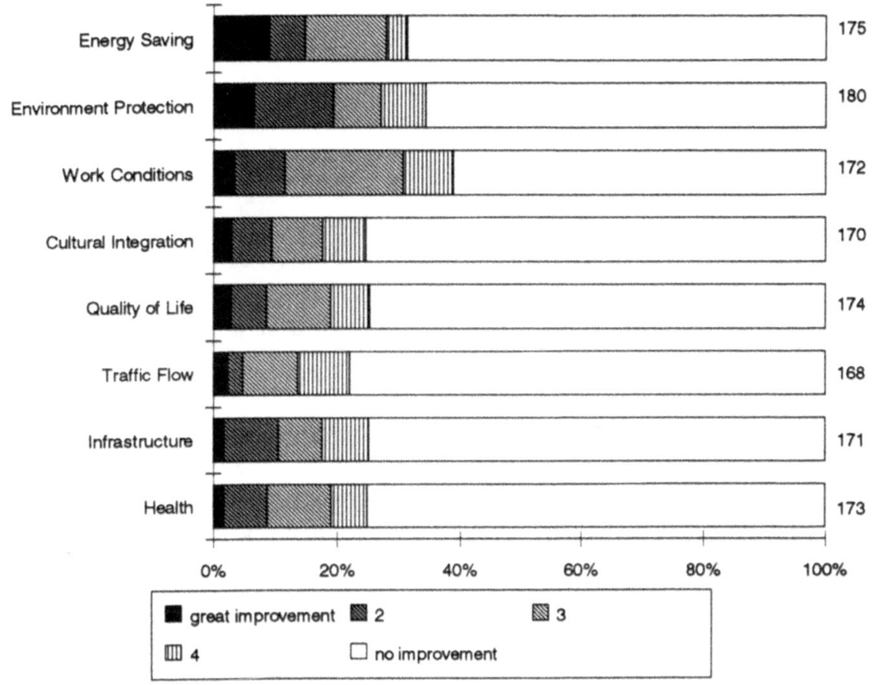

4.6 Importance of EC RTD Policy for Non-participating Enterprises

Potential Future Participants

Of the 277 firms answering our questionnaire that have not yet participated in research support by the EC, **90.3% conduct in-house research and development** and could thus be potential participants. The proportion of SMEs (89.9%) is somewhat smaller here than the proportion of BIGs (98.1%). For one quarter of enterprises (42.0% of BIGs, 21.3% of SMEs), **expenditure on internal R&D** constitutes between 1.5% and 3.4% of their total expenditure, for one third (36.0% of BIGs and 31.4% of SMEs) it

3.4% of their total expenditure, for one third (36.0% of BIGs and 31.4% of SMEs) it amounts to between 3.5% and 8.4% and for just under one quarter (6.0% of BIGs and 27.1% of SMEs) it is more than 8.5%. BIGs conduct their R&D mainly in their separate departments or in a central R&D unit; however, according to replies to the questionnaire only one quarter of SMEs have organisationally defined R&D departments, and most of them carry out R&D cross-departmentally in the manufacturing unit or in a project team. The R&D emphasis in firms replying to the questionnaire is on applied research and development activities. Viewed in its entirety, the sample represents a **realistic potential of firms** which might be adressed by the research support of the CEC, according to the firms' innovation behaviour.

Cooperation in the area of R&D also plays an important role for the non-participating group of firms; 75.9% of the SMEs in the sample and 96.2% of the BIGs are cooperating with other institutions in the area of R&D (either with formal contracts or as informal cooperation). However, this cooperation is **was up to now limited to partners in Germany**.

More than half of the non-participating enterprises (55.2%) intend to **participate in an R&D project supported by the CEC** within the next 5 years; this intention was expressed equally by large companies and SMEs. Approximately one fifth of firms (20.9%) will not be participating in EC RTD support because they wish to take part in national research support programmes only. 11% of firms replying to the questionnaire rejected the idea of publicly supported RTD programmes in general.

If firms join an EC project consortium, over half of them wish to participate as **contractors**. One third of firms replying to the questionnaire (33.1%) would aim to take on the position of main contractor; amazingly, it is above all the SMEs in the sample who feel confident to take this on.

Barriers and Difficulties

What are the reasons and barriers that have prevented this group of firms from participating in EC projects so far? The **main barriers to participation** are to be found the following four points (see Figure 4.6.1):

1) Deficits in information about the possibilities of EC support,
2) Process of application too complex and time-consuming,
3) Administrative and organisational tasks take up too much time and money,
4) EC projects go on too long (3-4 years).

The **complexity and expense** (in terms of time and money) **of the application process** tend to scare off especially **SMEs** answering the EC calls for proposals. This conclusion is also confirmed by interviews with experts: small and medium-sized firms **lack the resources** needed in order to release personnel to make the application, attend the "Proposers' Days" or the "Technological Days" organised by the EC in Brussels, or to search for suitable partners and reach agreements with them. A particular difficulty for SMEs when applying is that often they are not familiar with either the requirements or the assessment criteria for a successful application. Public advisory offices can give support by providing basic information, but this is not sufficient to overcome concrete impediments in application. For this aspect the firms interviewed were strongly in favour of **intensive, individual consulting**.

Deficits in information on the possibilities of EC RTD support do not represent a barrier only for SMEs. In some interviews with large, export-oriented enterprises in the mechanical engineering sector, we ascertained that even in these companies it was not known that the European Community also support research and technological development.

Lack of in-company qualifications was described on average by half the firms replying to the questionnaire, not as a serious but as a medium barrier to participation in EC RTD support. This refers primarily to a lack of experience in international project management among the personnel, but also to a lack of knowledge of other languages and an insufficient number of skilled, specialist personnel (cf. Figure 4.6.1).

Another factor regarded by non-participating firms as a barrier to participation in EC projects is the **low rate of approval of project applications**. These rates vary from programme to programme: for instance, in BRITE/EURAM (research topics I to IV) about (28.2%) of project proposals were approved (Commission of the EC 1993a), in the DRIVE programme 1 project in 3 was accepted (32.7%) (Commission of the EC 1990c).

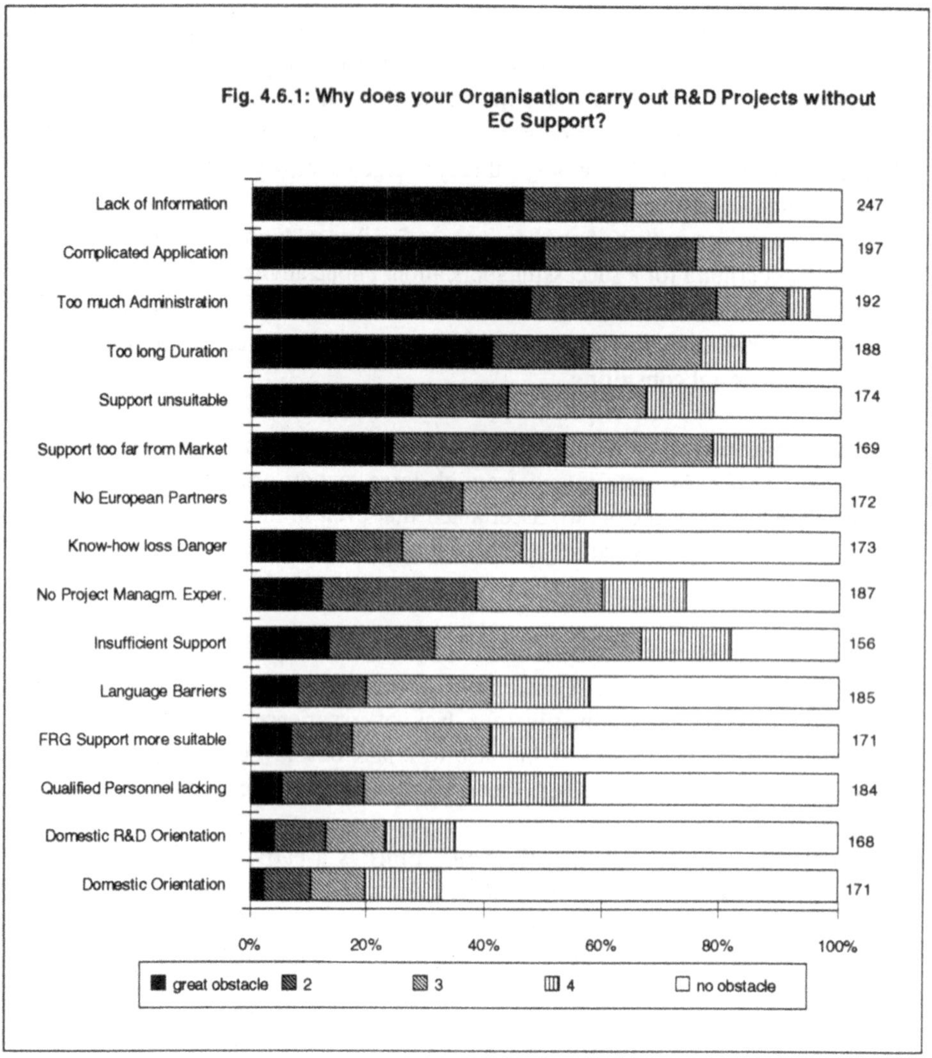

4.7 The Future Development of National and Community Policy

In the first place, **non-participating enterprises** consider it (very) important to **extend German R&D support**. Only in the second place do they wish for an extension of the **research support of the EC** (cf. Figure 4.7.1). If EC research support were to be strengthened, this should be primarily in the fields of **application-oriented research** and

market introduction of new products. From the viewpoint of non-participating firms, the areas of basic research and technology transfer have only a medium significance for the future orientation of EC actions. The promotion of cooperations outside R&D is not assessed as a priority by non-participating firms.

Fig. 4.7.1: The Relations of R&D Support by German Agencies to the EC will continue to change. What do you consider important?

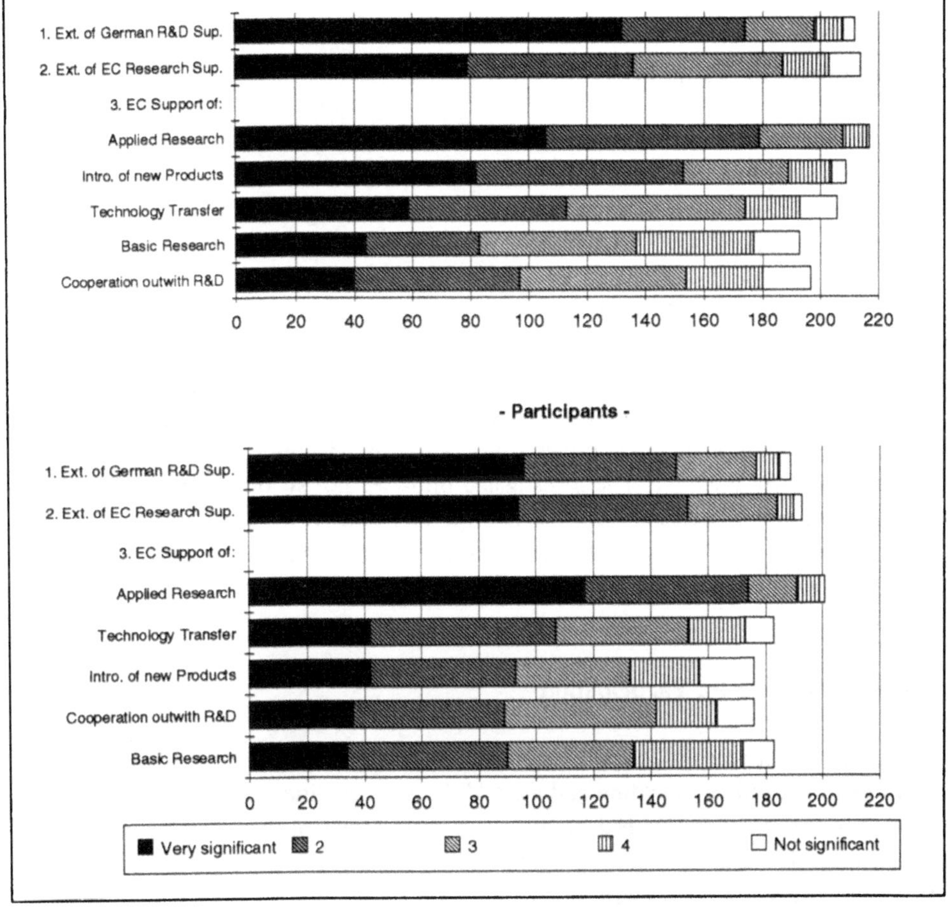

The **firms which have already participated in EC R&D projects** consider the **extension of the EC research support and the German R&D support** to be **equally** very important, unlike the non-participants (see Figure 4.7.1). The EC Commission should extend support above all in the fields of **application-oriented research, tech-**

nology transfer and then in third place - in contrast to the non-participants - the market introduction of new products. The support for basic research is also not considered an urgent priority, and the same applies for the promotion of cooperation relations outside the area of R&D.

4.8 Summary and Conclusions

Importance of EC Projects

From the viewpoint of participating firms, the R&D activity of EC projects can be characterised as **applied research** and **medium-term oriented** concerning the time horizons. The R&D activities of large companies are more long-term oriented and those of SMEs short-term oriented.

EC projects have a **definite strategic importance** in the **R&D portfolios** and in the **existing core areas** of firms; they serve to expand existing fields of business, and to a limited extent to establish new business fields or bring about diversification. Their importance for **technological competence** is assessed ambivalently by BIGs: according to the results of the questionnaire, EC projects have a **limited strategic importance for their technological competence**. The investigated firms see **not so much an improvement in their technology position relative to Japanese and American competitors** as an improvement in their competitiveness relative to firms in Germany and in other EC Member States.

Aims and Reasons for Participation

The predominant reasons for participation in EC RTD programmes are the **improvement of European cooperation** and the additional possibilities for financing R&D projects. Another reason mentioned by BIGs is that the topics of the EC RTD programme complement the German support programmes; SMEs feel that EC projects enable them to give their R&D a longer-term orientation. Among the project aims, it is "soft" aims that occupy the foreground, such as **increasing the knowledge base** and achieving **better cooperation with European firms**; improvement or innovative development of products ranks only second or third, depending on the EC programme line.

Impacts on Firms

The impacts of EC RTD support on enterprises also correspond to the primary reasons for, and aims of participation. Firstly, EC projects are **technically, scientifically and strategically useful** to firms (e.g. new fields of research, changes in technology strategy), and either stabilise or support their **cooperation with European firms**.

By contrast, the **economic impacts** of EC RTD support, such as increased turnover and larger market shares, are **still small** - and where these have occurred, the increase is relative to the EC market and not to the two main competitors, Japan and the USA. Due to the "time-lags" that elapse between project results, their incorporation in innovative or improved products, and the market introduction of these products, it has not been possible to establish **empirically** the economic benefits of the EC projects.

Regarded as a whole, the results of the questionnaire addressed to German firms show that EC projects increase scientific capabilities and influence technology strategies. By **stimulation of the scientific and technological basis** - at least for a considerable proportion of participating enterprises - **an important aim of the Second Framework Programme has been achieved**.

Participation in EC consortia **stabilises and reinforces cooperation** at a European level; this applies particularly to cooperation between enterprises. EC RTD support has less of an initiation function for cooperations: the majority of firms have already worked in numerous cooperations, and at the call for proposal applications are often made by complete consortia (or consortium "cores") which have already cooperated previously. EC RTD support offers SMEs an opportunity to expand their European business relationships to include cooperations in the area of R&D, thus fulfilling an important function for these firms.

There are rarely **social or environmentally relevant impacts** for the enterprises participating in EC projects, unless the project aims are expressly oriented towards them.

Potential Future Participants in EC RTD Support

More than half of the group of non-participating firms (55.2%) intend to take part in EC RTD support within the next five years, and are clearly in a position to participate as regards their innovative behaviour; thus **there is a potential of enterprises in Germany for future participation in EC RTD programmes.** The most important **barriers** for non-participants are deficits in information about the possibilities of EC support, problems in applying, and the expectation of an expensive and time-consuming administration procedure. The low rate of acceptance for proposals, which varies from programme to programme, is seen as another barrier.

Future Development of RTD Support

Unlike non-participants, enterprises that have already taken part in EC R&D projects assess the **strengthening of EC R&D support and of German R&D support equally** as very important. They would welcome an extension of support first of all in the fields of **applications-oriented research,** which is further away from the market and competition. The support of technology transfer and the introduction of new products is named as less significant. The support of basic research is not considered as an urgent priority either, nor is the support of cooperation relations outside the area of R&D.

5. Importance of EC RTD Policy and its Influence on Institutes of Higher Education

The German higher education sector is regarded as well developed and - following a wave of foundations in the 1960s and 1970s - well distributed regionally. In the new Länder (previously the GDR) the higher education system was totally re-organised at the beginning of the 90s (see also Krull 1992). Today there are 312 higher education institutes **(HEIs)** in Germany (state at the beginning of 1992); these comprise

- **86 academic universities** ("Hochschulen") that are universities with the right to confer PhDs. With a few exceptions, the universities engage in teaching and research in a wide range of scientific disciplines. After the research laboratories of industry, the universities represent the most extensive research potential in Germany.

- **124 technical colleges** ("Fachhochschulen"): These arose in the 1970s (usually superseding previous, older institutions, such as state colleges of engineering) and provide practice-oriented, job-oriented, science-based training particularly in the various fields of engineering, especially electrotechnics, mechanical engineering, civil engineering and architecture. The technical colleges do not have the right to confer PhDs. For some years now, they have increasingly been performing applications-related R&D, and they are also involved in the process of technology transfer to regional industry. For these reasons they were included in this survey.

- About **100 other higher education establishments** in the fields of teacher training, theology, music, the fine arts and public administration; these are not included in this survey.

The higher education sector has almost **1.8 million students** (1991). In the old Länder there are **103,000 academics** employed in the **sciences and the arts**, including 44,000 researchers. In the new Länder there are 79,000 employees in higher education, of whom 19,000 are researchers (statistics for 1991; cf. BMFT 1993). In the meantime, the numbers of higher education employees in the new Länder have gone down considerably.

The analysis in this chapter is primarily concerned with the **research activities of the research groups and institutes of the universities**. The technical colleges, because of their stronger, more direct orientation towards applications and their less extensive basic equipment for research (personnel, laboratories and apparatus) participate much less in "precompetitive" research; thus they do not figure so prominently in this analysis.

The **funding of research** in German higher education institutions has a **dual basis** (cf. Wissenschaftsrat 1988):

- The higher education budget finances the **basic equipping** with staff, premises, apparatus and non-cash resources. It covers research, education of next generation scientists and teaching functions, as well as smaller research projects and the preparations for larger research projects.

- For larger research projects, individual academics or institutes receive supplementary funding from a third party. This **"third party funding"** has to be acquired by the scientists themselves. It comes mainly from various public budgets and is granted for a limited time only.

Recently, this "dual principle" in research support is being supplemented in an increasing number of Länder by **additional research funding from the Länder**.

According to estimates by the **German Science Council** (Wissenschaftsrat 1988, 36), two thirds of expenditure by HEIs on research and development (R&D) in 1985 was financed by basic funding (including investments). **One third of R&D expenditure** in higher education institutions was financed by **third party funding**, 80% of which came from public and 20% from private sources including foundations. In many countries, the level of third party funding is higher (data for 1982): Germany 21%, Japan 25%, Great Britain 28%, France 48%, USA 77% (cf. Martin/Irvine 1986).

Third party funding in Germany has gone up in real terms by a quarter since 1975, whereas the basic funding used for research and teaching has remained more or less constant. Since a growing proportion of the basic funding has had to be spent on teaching, research support from basic funding has declined. Thus **third party funding** has become **increasingly important for research** in German HEIs (cf. Wissenschaftsrat 1988, 36). Third party funding is concentrated - corresponding to the relative costs of the research - in the **natural and engineering sciences.**

The most important source of third party funding for higher education is the **German Research Association** (Deutsche Forschungsgemeinschaft, **DFG**). In 1985, **42% of all third party funding**[29] in higher education institutions came from the **German Research Association** (DFG), which supports both individual projects and research programmes with a medium-term orientation (**special research areas/**"Sonderforschungsbereiche").

---

29 Based on Wissenschaftsrat 1988, 37

In the last few years the Federal Government has become increasingly important as a source of research funds for HEIs (1985: **21% of all third party funding**). In particular the Federal Ministry for Research and Technology (BMFT) is financing an increasing proportion of research in higher education institutes with its specialist programmes.

The importance of **industry** for research support has also grown: in 1985, **15% of all third party funding** came from this source. In some technically oriented HEIs the share of research financing from industry is considerably above this level.

According to **more recent statistics** (Wissenschaftsrat 1993, 56f), in 1990 48% of third party financing from **public sources** came from the DFG and 38% from Federal Ministries; the share of **industry** has increased only slightly since the 1980s.

The structure of third party funding in HEIs shows **regional deviations** from this pattern; for instance, the breakdown of third party funding in the universities of Baden-Württemberg in 1992 is as follows:[30]

- 27.1% from federal sources,
- 3.2% from the "Land" Baden-Württemberg
- 37.9% funding from DFG,
- 27.3% from industry,
- 4.5% EC funding.

Against this background, the financing of research in German institutes of higher education by the **European Community** has **gained ground** only **slowly**. Whereas EC funding was hardly perceptible in 1985, its share five years later was just under 2% as a proportion of research financing by HEIs, including basic funding (more precise data were not available up to now; see also Figure 2.1.1 in Chapter 2).

5.1 Characteristics of Participants and Non-participants in the Survey

Our questionnaire was answered by 201 research groups and institutes in higher education that had participated in the Second Framework Programme.

The research groups or institutes have an average size of less than 50 scientists (full time equivalents). The participants employ a larger average scientific staff than the non-par-

30 Data according to information from the Ministry for Science and Research of Baden-Württemberg, Stuttgart, June 1993

ticipants (cf. Table A 5.1.1 in the Annex). The total budget of the institutions investigated is relatively low in the majority of cases; however, participants have a higher average budget than non-participants.

HEIs taking part in EC RTD programmes show regional concentrations in the following "Regierungsbezirke" (primary administrative divisions of the "Länder"): Köln (11.4%), Braunschweig (9.5%), Arnsberg (7.0%) and Oberbayern (6.5%). Because of the fixed times of the EC calls for projects, only two HEIs in the new Länder were able to take part in EC projects following German reunification (see Table A 5.1.2 in the Annex, and also Chapter 7).

Most EC participants (90%) receive funds from other public RTD programmes. However, three quarters of EC non-participants also receive support from these programmes. Table 5.1.1 shows where this funding comes from: **70% of EC participants** and **70% of non-participants** are involved in **BMFT programmes**. By contrast, industrially-oriented EUREKA projects and support programmes of the BMWi (Federal Ministry for Economic Affairs) are of secondary importance for the HEIs. Almost half of the non-participants (47.5%) and participants (49.7%) questioned receive funds from other bodies (in particular from the DFG, from foundations and from industry). Approximately one third of the HEIs questioned have had support from RTD programmes of the Länder. Thus the differences between participants and non-participants (insofar as they receive support at all) are small: it is mainly the **EC funding** itself which **constitutes the difference** between the two groups.

Table 5.1.1: Is your Organisation Participating in (Other) Publicly Supported RTD Programmes?

RTD programmes	Participants		Non-participants	
	absolute	%	absolute	%
No	22	10.9	44	23.9
Yes, namely:	179	89.1	140	76.1
further EC projects	97	54.2		
Eureka projects	17	9.5	8	5.7
BMFT support programmes	125	69.8	99	70.2
BMWi support programmes	21	11.7	19	13.5
Länder programmes	60	33.5	50	35.5
other programmes	89	49.7	67	47.5

Multiple answers were possible.

5.2 Description of Sample and Comparison with Basic Population

In order to assess the representativeness of the questionnaire answered by participating higher education institutes, it is necessary to compare the characteristics of the sample with those of the basic population. The share of HEIs responding to the questionnaire corresponds to the structure of German participants in the Second Framework Programme. Institutes of higher education make up approximately one third of all German participants analysed.

There is also correspondence between the structure of German participants and the sample regarding the position of HEIs in the project consortia. Contractors (59.1%) are slightly over-represented in the sample compared to the general population (52.1%), with the share of associates (12.2%) and subcontractors (8.1%) being slightly higher for the population as a whole (cf. Table 5.2.1). More than one quarter of HEIs in the sample (29.8%) participate in EC projects as **main contractors**; this reflects the strong participation of these institutions in programmes such as SCIENCE. Some interviewees stated that HEIs mostly cannot take on the role of project coordinator in large, industrially-oriented projects, due to the great amount of organisation this involves.

Table 5.2.1: Is your Position in Project that of...?

	HEIs Survey Sample		HEIs Basic Population	
	absolute	%	absolute	%
Main contractor	59	29.8	184	27.1
Partners	117	59.1	354	52.1
Associates	14	7.1	83	12.2
Subcontractors	8	4.0	55	8.1
Total	198	100.0	676	99.4

The distribution of participation in the various EC RTD programmes in the sample and in the basic population generally corresponds, with three exceptions: BRITE/EURAM and STEP/EPOCH are over-represented in the answers to the questionnaire, whereas ESPRIT is under-represented (cf. Table 5.2.2).

Table 5.2.2 Structure of the answering HEIs in EC Programmes

EC Programmes (Second Framework Programmes 1987-1991)	Universities (n=201)		Parent Population (n=679)	
	absolute	in %	absolute	in %
1. Quality of Life	**31**	**15.4%**	**73**	**10.8%**
e.g.: Human Gen. Analysis	1	0.5%	0	0.0%
Medicine/Health	2	1.0%	8	1.2%
Radiation Protection	1	0.5%	18	2.7%
STEP/EPOCH	27	13.4%	47	6.9%
2. Information and Communication Technologies	**37**	**18.4%**	**232**	**34.2%**
e.g.: AIM	3	1.5%	12	1.8%
DELTA	1	0.5%	2	0.3%
DRIVE	6	3.0%	30	4.4%
EURET	0	0.0%	0	
ESPRIT II	26	12.9%	182	26.8%
RACE I	1	0.5%	6	0.9%
3. Modernisation of Industry	**32**	**15.9%**	**75**	**11.0%**
e.g.: B.C.R	1	0.5%	0	
BRITE/EURAM	24	11.9%	53	7.8%
Raw Materials	7	3.5%	22	3.2%
4. Biological Resources	**8**	**4.0%**	**21**	**3.1%**
e.g.: Agriculture	0	0.0%	0	
BIO/BRIDGE	6	3.0%	16	2.4%
ECLAIR	2	1.0%	5	0.7%
FLAIR	0	0.0%	0	0.0%
5. Energy	**27**	**13.4%**	**68**	**10.0%**
e.g.: Decommissioning	1	0.5%	6	0.9%
Fusion	0	0.0%	0	
JOULE	23	11.4%	59	8.7%
Radioactive Waste	2	1.0%	2	0.3%
TELEMAN	1	0.5%	1	0.1%
6. Science and Technology for Development	**10**	**5.0%**	**24**	**3.5%**
7. Marine Resources	**4**	**2.0%**	**17**	**2.5%**
e.g.: FAR	0	0.0%	0	
MAST	4	2.0%	17	2.5%
8. Improvement of European S/T Cooperation	**44**	**21.9%**	**169**	**24.9%**
e.g.: DOSES	0	0.0%	0	
EUROTRA II	0	0.0%	3	0.4%
Large-scale facilities	0	0.0%	0	0.0%
MONITOR	0	0.0%	3	0.4%
SCIENCE	44	21.9%	155	22.8%
SPES	0	0.0%	6	0.9%
VALUE	0	0.0%	2	0.3%
9. Other (related) Programmes	**8**	**4.0%**	**0**	**0.0%**
Total	**201**	**100.0%**	**679**	**100.0%**

In the following sections, the answers of the institutes questioned are grouped and evaluated, wherever appropriate, in three EC programme lines (corresponding to the programme lines of the Second Framework Programme) and one combined group, as follows:

- The scientifically oriented programme line "**Improvement of European Science and Technology Cooperation" (S&T Cooperation)** accounts for 22% of answers, with the emphasis on SCIENCE.
- The programme line "**Information and Communication Technologies" (I&C Technologies)** covers 18% of answers, with clear emphasis on ESPRIT and to a lesser extent on DRIVE.
- The programme line "**Modernisation of Industry**" covers 16% of answers, mainly reflecting BRITE/EURAM participation.
- The rest of the answers relate to participation in various different programmes (particularly STEP/EPOCH, JOULE); because of the small numbers involved, these have not been analysed separately but have been put together in an "**other areas**" group.

5.3 Importance and Strategic Role for Participating Institutes

5.3.1 Characterisation of EC Projects

The higher education institutes answering the questionnaire often participate in EC projects with a relatively large number of partners (average 6.8 partners) (cf. Table 5.3.1.1). They tend to participate particularly in projects in which other HEIs are strongly represented; they are obviously **seeking similar partners**. They clearly tend to cooperate less with other types of R&D performers - of these, they cooperate most often with enterprises.

More than half (54.1%) of the EC projects carried out by HEIs have a **long-term orientation** and exploitation of the results is envisaged only three years after the end of the project (cf. Table 5.3.2.1 in section 5.3.2). Unlike firms or application-oriented R&D institutes, the universities are interested in long-term R&D activities and do not have to rely on the economic exploitation of results.

Table 5.3.1.1: How many Partners and Associates have participated in this EC Project?		
	HEIs	
	M	**se**
Firms from Germany	0.4	0.8
Firms from EC countries	1.1	0.2
Universities from Germany	1.3	4.9
Universities from EC countries	2.5	4.1
R&D Institutes from Germany	0.2	0.5
R&D Institutes from EC countries	0.9	1.8
Other organisations from Germany	0.0	0.1
Other organisations from EC countries	0.1	0.7
Total	6.8	8.1
M = Mean; se = standard error		

Taking projects in which HEIs participated as a whole, the main emphasis is on **basic research** (45.3%) and **applied research** (42.3%). By contrast, experimental development and demonstration projects/diffusion play only a marginal role. The **majority of own project tasks** performed by the research groups and institutes questioned were in **basic research (50,8%)**, a higher proportion than for the projects as a whole (cf. Table 5.3.1.2), whereas applied research (40.2%) has a less significant position. This result emphasizes the concentration of universities on the field of basic research; however, **applied research** plays ***more* than just a minor role.**

Table 5.3.1.2: Which Type Characterises Best Your Share in the Whole Project ?		
	HEIs	
	absolute	**%**
Basic research	101	50.8
Applied research	80	40.2
Experimental development	12	6.0
Demonstrators and diffusion	6	3.0
Total	199	100.0

5.3.2 Strategic Role of EC Projects

Most of the projects carried out are an **integral part of the R&D portfolio** of higher education institutes (76.4%). Only a small proportion of projects have no connections to other R&D projects (cf. Table 5.3.2.1). This result shows that the HEIs tend to participate in EC promotion if the cooperation lies in their established fields of research and they are able to perform the appropriate research work.

Table 5.3.2.1: Composition of the R&D Portfolio

	HEIs	
	absolute	%
What Connection does the EC Project have to other R&D Projects in your Organisation?		
Integrated part of R&D portfolio	146	76.4
Few connections to other R&D projects	31	16.2
No connection to other R&D projects	14	7.4
Total	191	100.0
Time span of this Project		
Short-term	20	12.1
Medium-term	60	36.4
Long-term	85	51.5
Total	165	100.0
What Role does the EC Project play for your technological Competence (technical standing)?		
Strategic core area	93	52.2
Potential new core area	27	15.2
Just one technology among others	50	28.1
Rather peripheral technology area	8	4.5
Total	178	100.0
What Role does the EC Project play in your Strategy?		
Establish new research areas	20	14.2
Extension of a research area	104	73.8
No significance	5	3.5
Diversification	9	6.4
Cost reduction	3	2.1
Total	141	100.0

The majority of EC projects (52.2%) are carried out by HEIs in their **established strategic core areas**. Only 15% of EC participants intend to build up a potential new core area with the help of the projects. In a few cases (4.5%) the research field of the EC project is only in a peripheral technology area (cf. Table 5.3.2.1).

It is clear that by participating, HEIs are not so much trying to open up new fields of research (the equivalent of a new "business field" for enterprises), since only 14.2% of them attribute this function to EC projects. Their participation is definitely directed primarily towards **extending existing research areas** (73.8%). However, it emerges in replies to questions on the impacts of participation in EC projects (see section 5.5.1) that - especially in projects in the science-oriented programme line "S&T Cooperation" - new research fields have often been entered.

More than 60% of HEIs state in their replies that the project could not have been carried out without EC support (cf. Table 5.3.2.2); in one third of cases, EC support has enabled the project to be carried out on a larger scale. Only 4% of replies stated that the project would have been carried out independently of EC funding. Thus it is clear that **EC funding** has **initiated** original, **additional research activities**. However, the research was not accelerated by EC support; only 18% of institutes' replies indicate that this was important to them. Since HEIs are concerned mainly with basic research and have long term perspectives, it is hardly surprising that the time element (i.e. faster research) is only of secondary importance to them.

Table 5.3.2.2: What Role did the EC Funding Play?

	HEIs	
	absolute	%
Independent of funding	8	4.0
Faster through funding	36	18.2
Larger scale possible through funding	59	29.8
Other goals without funding	21	10.6
Not at all without funding	121	61.1
Total	198	100.0

Multiple answers were possible.

Table 5.3.2.3 shows that as a **source of financing, EC funds** (17% of research budget financing) already play a much more **important role** today in the research budgets of participating HEIs than indicated by the average statistical values given in section 5.1. Nearly two-thirds of the HEIs investigated (63.5%) expect the EC to become

Table 5.3.2.3: What Share in the Financing of your R&D Budgets do the following have?

	Mean 1991 in %	in future decreasing		in future constant		in future increasing		Total	
		abs.	%	abs.	%	abs.	%	abs.	%
Own funds	23.9	45	33.1	88	64.7	3	2.2	136	100.0
Own contracts	0.1	1	6.3	13	81.3	2	12.5	16	100.0
Contracts from firms	8.2	2	2.9	29	42.0	38	55.1	69	100.0
Contracts from HEIs and RDIs	0.5			22	78.6	6	21.4	28	100.0
EC Funding	16.8	17	11.5	37	25.0	94	63.5	148	100.0
Federal Republic funding	27.6	50	36.2	61	44.2	27	19.6	138	100.0
Länder funding	10.0	29	31.5	41	44.6	22	23.9	92	100.0
Other sources	12.2	8	9.3	44	51.2	34	39.5	86	100.0
Total	100.0								

a more important source of financing their R&D budget in future. On the other hand, it is feared that the share of funding from the Federal Government and the Länder may decrease.

5.4 Reasons and Aims of Participation, Results and Success Factors

The principal motivation of HEIs in participating in **EC promotion** is the hope of **additional sources of finance** (see Table 5.4.1); this was stated in three quarters (77.8%) of replies from research groups and institutes. Two-thirds (67.7%) are of the opinion that EC programmes are a necessary supplement to German support. More than half mentioned that **European cooperation** is **important** in their field of technology.

Table 5.4.1: Why are you carrying out R&D Projects with EC Funding?

	HEIs	
	absolute	%
Further financing source	154	77.8
Enables long-term orientation of research	71	35.9
Necessary supplement to German support	134	67.7
European cooperation is important	105	53.0
German programmes unsuitable	18	9.1
Exploration of new business areas	6	3.0
European cooperations outwith R&D	44	22.2
Other reasons	14	7.1
Total	198	100.0

Multiple answers were possible.

Building up relations in Europe outside R&D (for instance, by student exchanges) is only secondary and is regarded, so to speak, as a beneficial side-effect of EC programmes.

The **most important aim** for participation in EC programmes is **improving the knowledge base** (enhanced skills: 85% of the institutes). This result reflects the academic and scientific function of university institutions. Scientific reputation, which is a corollary of these activities, is named as a goal by 130 HEIs. Although participation in EC programmes is not yet in itself regarded as a distinction in scientific academic circles, there is evidently acceptance of the aim of enhancing reputation. Interviews confirm that, in the opinion of academic scientists, the scientific quality of the supported projects has improved. Other effects considered important are the **support of PhDs, cooperation with other European universities** and the qualification of R&D personnel. Market oriented aims are generally unimportant.

Figure 5.4.1 shows that, in a **comparison** of the three different **programme lines** (S&T Cooperations, Information and Communication Technologies; Modernisation of Industry) the **weighting** of the institutes' most important goals is **similar**: academic behaviour orientation clearly dominates over programme-specific features.

Replying to the question whether the envisaged aims had been attained, the great majority (89%) of higher education institutes stated that their aim of **enhancing scientific skills** had already been achieved by the end of the project. The projects were **also successful** in achieving better **cooperation** with other European R&D institutes and universities, enhancing **scientific reputation** and increasing the **qualification of R&D personnel**. The PhD theses were only fully completed three years after the end of the project: 54% of PhDs were already finished at the end of the project, a further 38% three years later; this corresponds to the usual time scale for writing a thesis.

If **aim attainment** is differentiated according to programme lines (see Figures 5.4.2 to 5.4.5), there is a recognisable **reflection** of the **different programme line orientations.** In the science-oriented programme line "S&T Cooperation", the dominant aim of enhancing scientific skills is followed by achievement of scientific reputation, cooperation with other European institutes, qualification of R&D personnel and the achievement of PhDs. The programme lines with a technological and industrial orientation have a similar pattern in their principal goals, but the secondary aims diverge considerably: in "I&C Technologies", success in increasing the numbers of R&D personnel, improvements in methods and instruments, and the successful design of prototypes play an important role. In the area "Modernisation of Industry", process innovations emerge as the most often mentioned secondary goal.

Some research groups and institutes experienced disappointments in their aim of participating in further EC projects: 25 of the 84 HEIs which had expressed this aim did not attain it. This was true particularly for the programme line **"I&C Technologies"**; the **high rejection rates** in this field due to insufficient availability of funds are well-known.

Fig. 5.4.1: Which of the following Goals did you aim for in this Project? (HEIs; absolute figures)

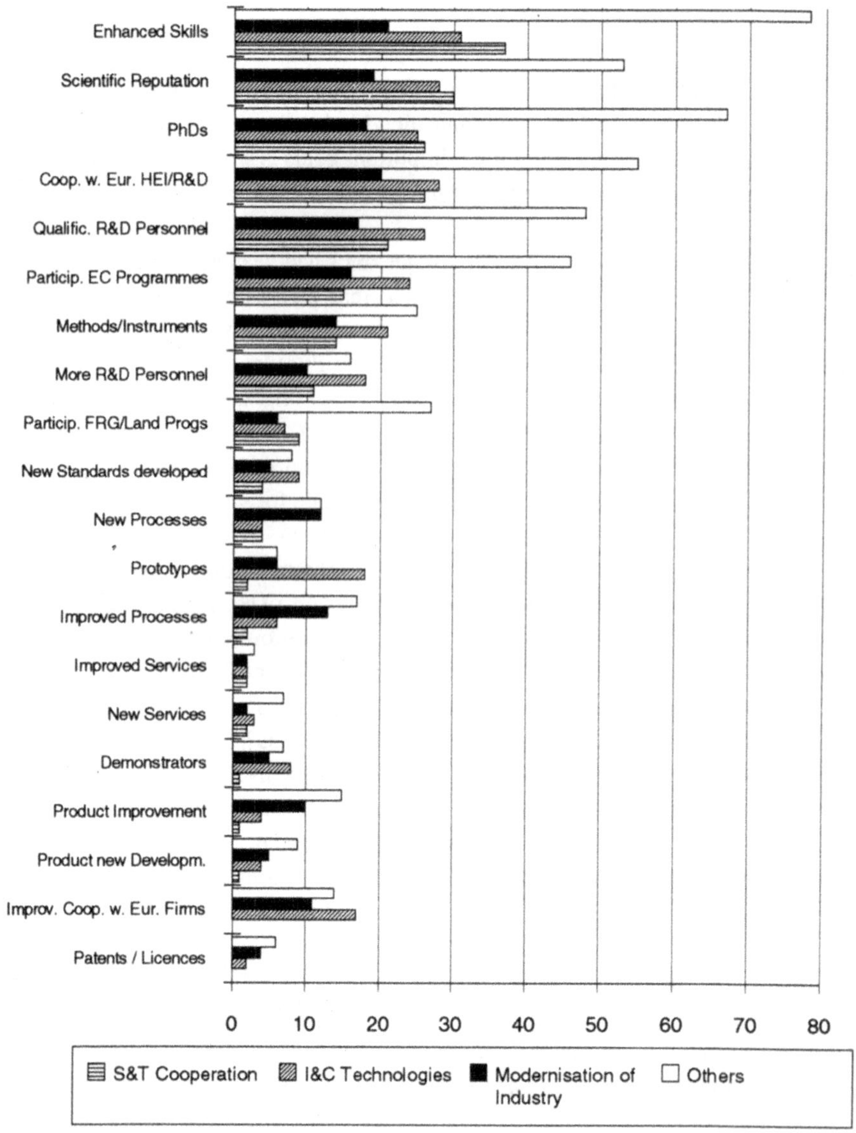

Fig. 5.4.2: Which Results did you achieve in this Project? (S&T Cooperation; HEIs)

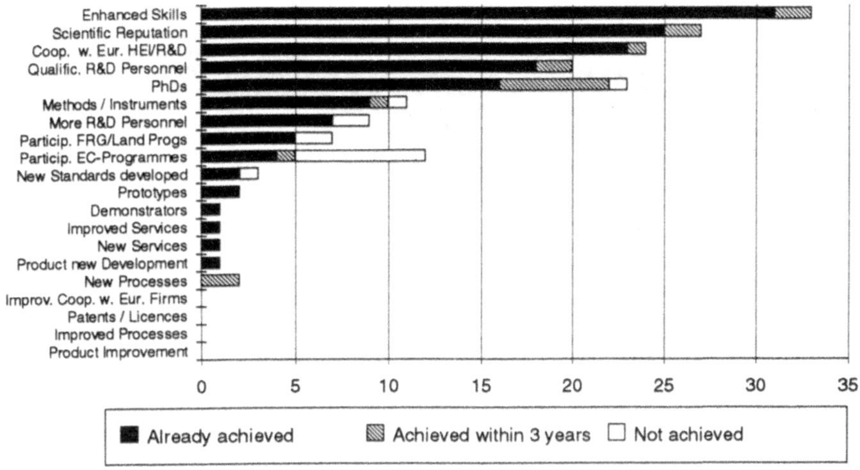

Fig. 5.4.3: Which Results did you achieve in this Project? (I&C Technologies, HEIs)

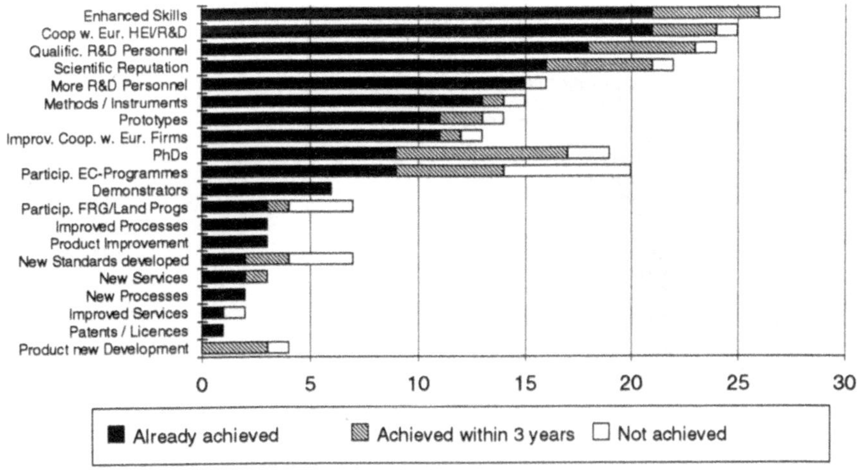

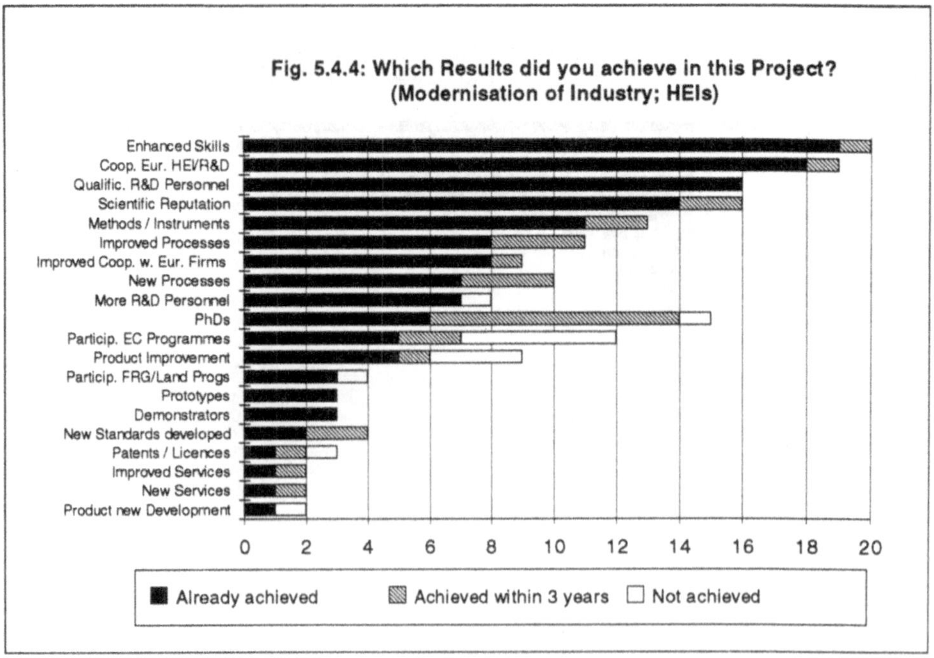
Fig. 5.4.4: Which Results did you achieve in this Project?
(Modernisation of Industry; HEIs)
Enhanced Skills
Coop. Eur. HEI/R&D
Qualific. R&D Personnel
Scientific Reputation
Methods / Instruments
Improved Processes
Improved Coop. w. Eur. Firms
New Processes
More R&D Personnel
PhDs
Particip. EC Programmes
Product Improvement
Particip. FRG/Land Progs
Prototypes
Demonstrators
New Standards developed
Patents / Licences
Improved Services
New Services
Product new Development
0
2
4
6
8
10
12
14
16
18
20
Already achieved
Achieved within 3 years
Not achieved

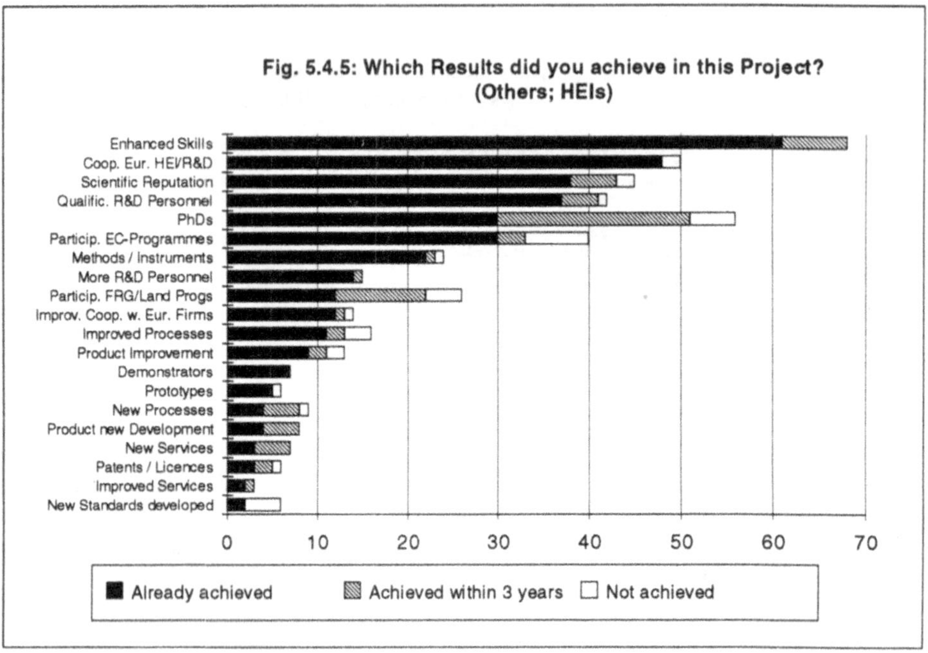
Fig. 5.4.5: Which Results did you achieve in this Project?
(Others; HEIs)
Enhanced Skills
Coop. Eur. HEI/R&D
Scientific Reputation
Qualific. R&D Personnel
PhDs
Particip. EC-Programmes
Methods / Instruments
More R&D Personnel
Particip. FRG/Land Progs
Improv. Coop. w. Eur. Firms
Improved Processes
Product Improvement
Demonstrators
Prototypes
New Processes
Product new Development
New Services
Patents / Licences
Improved Services
New Standards developed
0
10
20
30
40
50
60
70
Already achieved
Achieved within 3 years
Not achieved

In order to obtain a more detailed overview of written results emerging from the supported R&D projects, HEIs were asked about the numbers of their scientific publications. Compared with non-university R&D institutes (RDIs), there are **relatively fewer publications** by HEIs. One explanation for this may lie in the high standards applied in the academic sector to scientific publications.

Fig. 5.4.6: Which Factors have helped or hampered the Success of this EC Project or led to scrapping of Project?

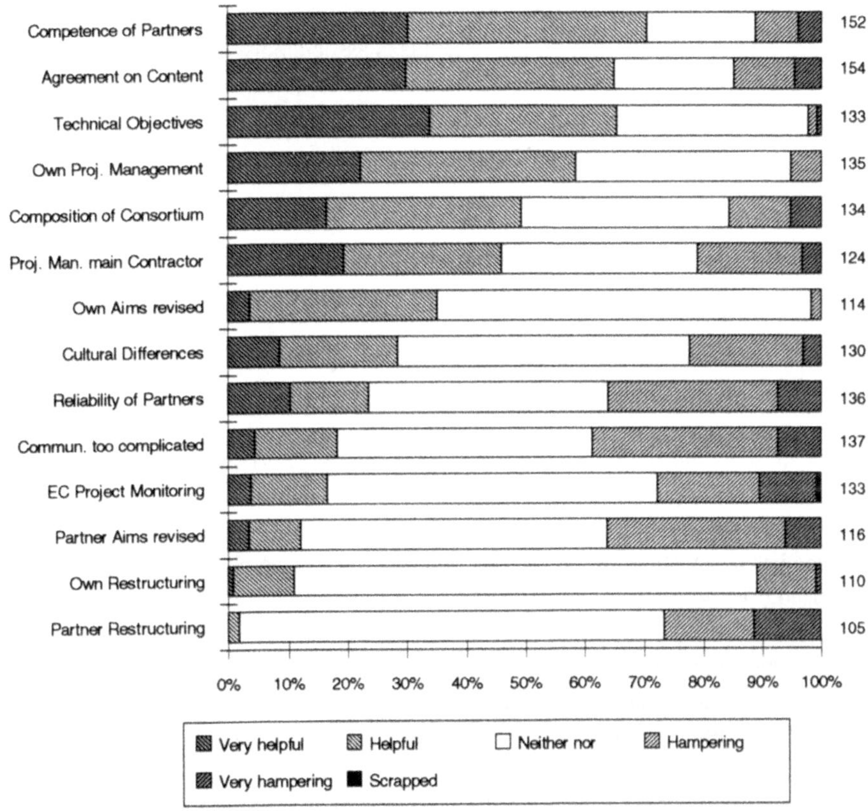

According to statements made in the questionnaire by HEIs, **factors** which have contributed to the **success** of the EC projects (cf. Figure 5.4.6) include:

- specialist competence of partners,
- agreement on the project's content,
- establishing of technical objectives,
- institute's own project management,
- composition of consortium.

Coordinated cooperation with partners in other countries within a well structured and effectively managed project is certainly the key to success in the supported projects. The **complexity of communication** in particular, and the failure of partners to stick to time schedules (lack of reliability of partners) remain **problematical**. The former factor was mentioned as an impediment in 40% of replies from HEIs. Differences in the culture and mentality of partners were considered by some HEIs as hampering, and by others as helping the success of the joint project. The tasks associated with project monitoring by the Commission were also considered in some cases as obstacles, and in other as positive for the success of the project.

5.5 Impacts on Participating Institutes

5.5.1 Economic, Technical and Scientific Impacts

It is difficult to evaluate the "business" impacts of EC support on higher education institutes, as they do not play a central role (cf. Figure 5.5.1.1 to 5.5.1.4). More than 70% of institutes replying to the questionnaire expressly emphasise that their **scientific performance** has substantially increased through participation in EC programmes. It is also emphasised (over 40% of replies) that

- EC support has contributed to establishing **new fields of activity** (R&D areas),
- activities in the area of **contract research** have **increased** (just under 20%),
- the **recruiting of scientists** from countries of the EC has been facilitated (over 15%) and
- it has been possible to **increase the numbers of R&D personnel** (ca. 15% of replies).

Fig. 5.5.1.1: Which Impacts has this EC Project had on your Organisation (S&T Cooperation; HEIs)

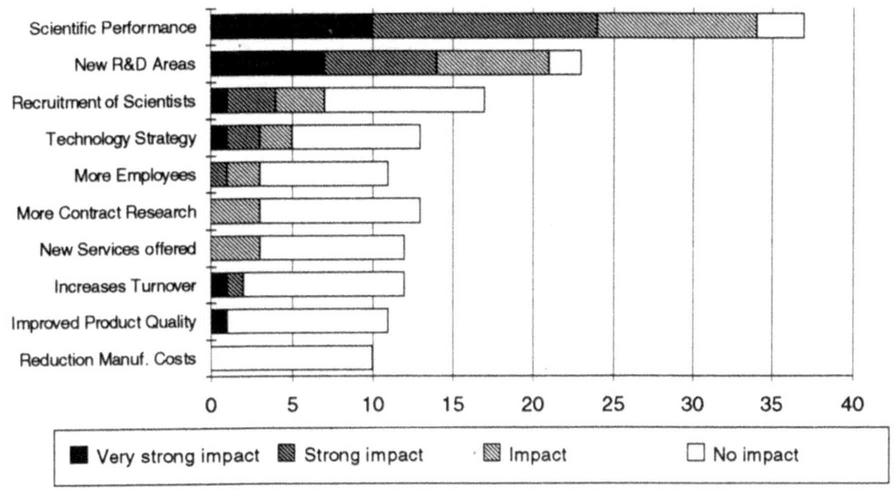

Fig. 5.5.1.2: Which Impacts has this EC Project had on your Organisation (I&C Technologies; HEIs)

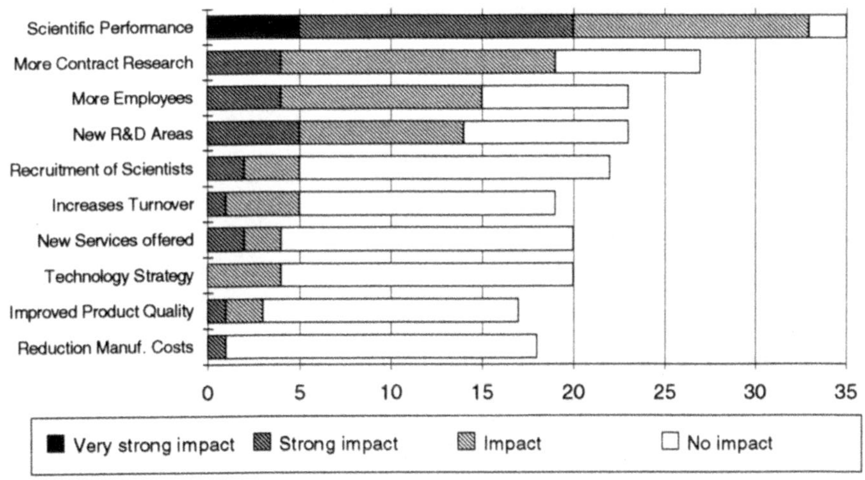

Fig. 5.5.1.3: Which Impacts has this EC Project had on your Organisation (Modernisation of Industry; HEIs)

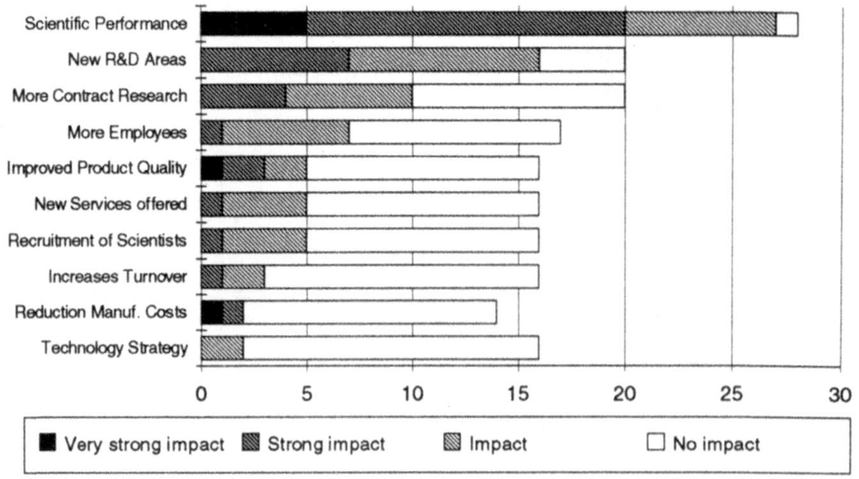

Fig. 5.5.1.4: Which Impacts has this EC Project had on your Organisation (Others; HEIs)

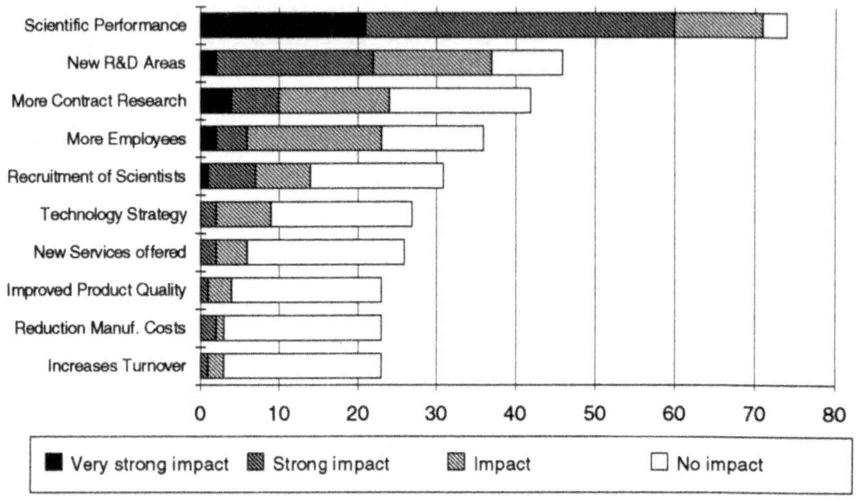

In a differentiated comparison according to **programme lines**, it is also an improvement in **scientific performance** that is the most frequently mentioned impact of EC projects. In the science-oriented programme line **"S&T Cooperation"**, the second most significant effect is the establishing of new research fields in the research group or institute; thus, project participation has initiated new perspectives. It has also become easier to **recruit science staff** from other EC countries; interviews with research groups and institutes show that this often takes place in the form of an exchange of students working for PhD theses or first degrees between partner institutes, who cooperate at times on other current projects.

In the programme line **"Modernisation of Industry"** (cf. Figure 5.5.1.3) higher education institutes also give **new research fields** as the second most important impact; some of them also mention an increase in contract research. More **contract research** and increasing the numbers of **R&D personnel** are the most sigificant secondary effects in the programme line **"I&C Technologies"** (cf. Figure 5.5.1.2).

5.5.2 Impacts on the Cooperation Behaviour

A large proportion of the higher education institutes analysed have already cooperated with external partners before participating in EC projects. The structure of cooperation partners is given in Table 5.5.2.1.

Table 5.5.2.1: Cooperation with other Organisations in R&D (Participating HEIs)

	Germany	EC	Rest of Europe	USA	Japan	Rest of world
Suppliers	6	3	3	2	1	1
Customers	10	5	3	3	1	1
Competitors	8	6	2	2	1	
Universities	127	124	82	84	32	49
R&D Institutes	69	62	23	30	11	18
Advisors	21	10	2	2		
Network	27	25	12	6	3	6

Multiple answers were possible

It is clearly apparent that the great majority (95%) of HEIs work with other HEIs, both in and outside EC projects. Apart from these, the preferred partners are non-university R&D institutes. There are relatively few contacts with enterprises. By contrast, HEIs are most frequently named as cooperation partners by enterprises and by RDIs.

Cooperations with other institutes of higher education are **worldwide**. Most importance attaches to **contacts within the European Community.** Before participating in the EC project, more than three quarters of the institutes in the survey had already cooperated with other HEIs in Germany and in the EC. Academic establishments in the United States are also interesting partners: more than half of the research groups that had participated in international cooperations had worked with higher education partners from the USA. Although relatively small, the number of cooperations with Japanese institutes is remarkable. Interview results reinforce the impression that the research groups and institutes that participate in EC projects are distinguished by an **above-average international orientation**.

Contacts with other types of partners (enterprises, RDIs) are usually concentrated inside Germany, and to some extent within the European Community. Once the EC borders are crossed, the number of contacts with non-university partners falls off.

In the search for partners for EC cooperations, **existing contacts** are the decisive **starting-point** (80%) (cf. Figure 5.5.2.1). **Previous cooperations** in other EC projects (a factor which will continue to gain importance as the numbers of participants increase), and the typical academic communication channels of conferences and publications, also play a significant part in the search for partners. External assistance from the EC and Federal/Länder sources, such as the EC information data Bank, the EC offices of scientific organisations or public contact agencies of the federal government and the Länder, were evaluated as having hardly any importance in the search for partners; however, it should be borne in mind that at the time when cooperations were being formed for the Second Framework Programme, these public contact agencies had either not yet been opened, or were still being built up. In the meantime, more and more HEIs are tending to install EC information offices or name officials in charge of European affairs; this emerges in the interviews. The external task of these offices or officials in charge is to make contact with the EC; their internal purpose is the diffusion of information and coordination of European activities within the university or technical college.

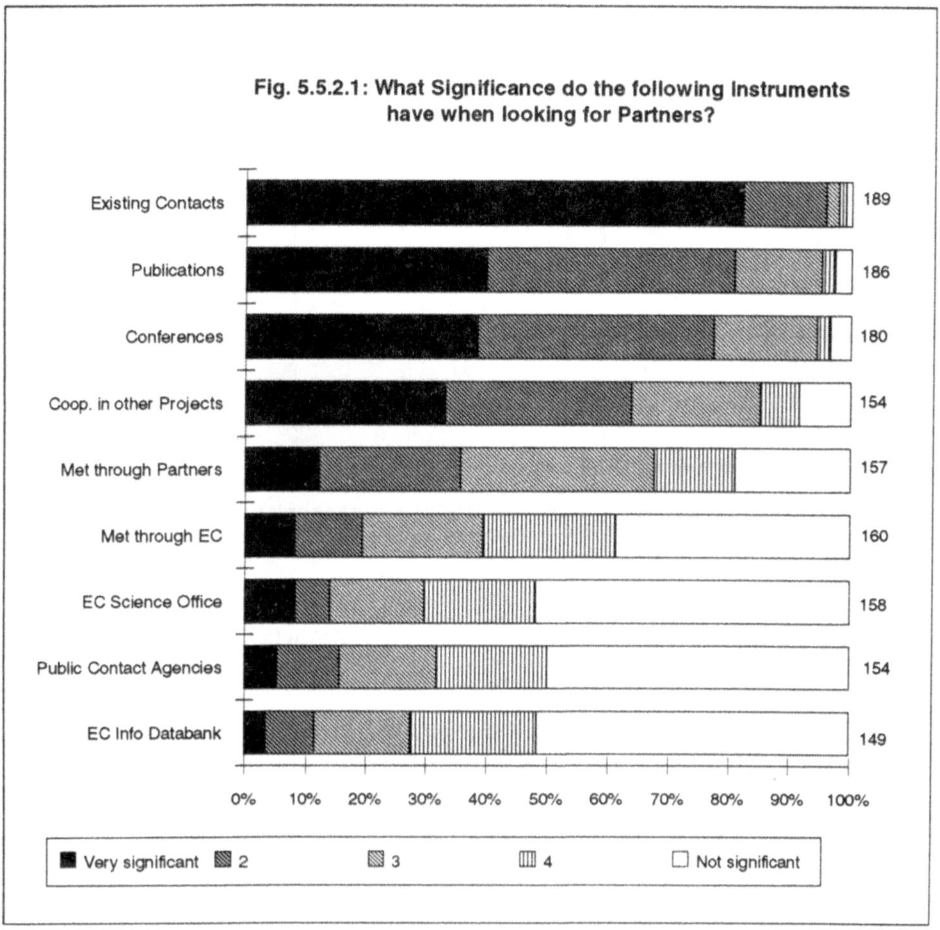

As a result of cooperation with European partners in projects promoted by the EC, the **willingness of HEIs to cooperate has increased markedly** (cf. Table 5.5.2.2). Almost 80% of research groups and institutes questioned state that their readiness to cooperate has increased following participation in the EC programmes. Cooperation in the area of R&D has been considerably extended (60% of HEIs), especially to include further EC projects. Interviews with academic researchers explicitly emphasise this. Experiencing the **possibilities of European research cooperation** is "irreversible", in the sense that cooperation horizons have been **permanently broadened**. Other studies also confirm this conclusion (cf. Tsipouri et al. 1992).

Table 5.5.2.2: What Impacts did your Cooperation with European Partners have?

	absolute	%
No further cooperations	6	3.2
Decreased willingness to cooperate	19	10.2
Increased willingness to cooperate	147	79.0
Extension of R&D cooperation	112	60.2
Extension of cooperation	15	8.1
Total	186	100.0

Multiple answers were possible.

5.5.3 Social and Environmentally Relevant Impacts

Most of the projects carried out by HEIs generally only have social aims to a limited extent (e.g. like improvement of working conditions or quality of life); correspondingly, impacts of this kind are described as small (cf. Figure A 5.5.3.1 in the Annex). According to approximately a third of participating institutes, EC projects have made improvements in environment protection. Perceptible impacts on energy conservation are also mentioned: one quarter of institutes replying to the questionnaire regard the supported project as contributing to cultural integration. Obviously, the social and environmentally relevant impacts are limited to those EC projects which explicitly follow the aims of social and environmental improvements.

5.6 Significance of EC RTD Policy for Non-participating Institutes

Almost all (99.5%) of the 189 HEIs that replied to our questionnaire as non-participants carry out their own research and development. Thus they at least fulfill the basic requirements for participation in EC RTD support.

Almost **one quarter** of the "non-participating" HEIs had already applied to an **EC support scheme**, but had been **turned down**. This proportion, which is higher than for enterprises or non-university R&D institutes, shows that HEIs have a **great readiness to participate** in projects, but that their success rate is low. The most significant cause for this is known to be the shortage of available EC funding. A further explanation may be that the academic research groups and institutes are more oriented towards basic research and therefore can commit themselves less to the industry-oriented EC programmes.

It is remarkable that in our sample the **non-participants** among the HEIs are more strongly **industry-oriented** than the EC participants (the experimental development and demonstrators/diffusion were more often mentioned by the non-participants than by the EC participants). The cooperation behaviour of non-participants emphasizes their stronger orientation towards industry; the cooperation with enterprises occurs on average more often than with participants. One reason may be that in EC project consortia, HEIs are in demand as suppliers of scientific competence rather than as industrially-oriented developers.

Non-participating higher education institutes clearly have **very effective networks within Germany**, but **far fewer contacts in other countries** (approximately half as many as participants). Existing contacts abroad are only to universities. Does this limited international orientation among non-participants provide an explanation for their non-participation in EC programmes, or has participation induced or accelerated an international orientation in the group of participants? The fact that EC participants' openness to contacts abroad also extends much more strongly to countries outside the EC, such as the USA and Japan for instance (see also section 5.5.2), refutes the second part of this question.

Why have the HEIs not yet participated in EC projects? The **main problems** are clearly to be found in the **application procedure**, which is thought to be **too complicated and time-consuming**, and in the complexity of the expected **administration** and **organisation** (cf. Figure 5.6.1): interviews show that there are also problems with contractual conditions and procedures. The EC is occasionally described by non-participants as a non-transparent bureaucracy which does not offer the security of the familiar procedures of the German support institutions (e.g. the DFG or the large foundations).

The following obstacles to participation are regarded as serious: the search for adequate European partners; the level of funding by the EC, which is considered to be too low (especially the 50% financing in "cost shared actions" with industrial partners is experienced as a problem); deficits in international programme management experience on the part of employees, and lack of information about EC support possibilities. Factors such as a danger of "know-how drains", lack of qualified specialist capacities, deficits in employees' language knowledge, or EC projects possibly lasting too long, are regarded as small obstacles only.

Fig. 5.6.1: Why does your Organisation carry out R&D Projects without EC Support?

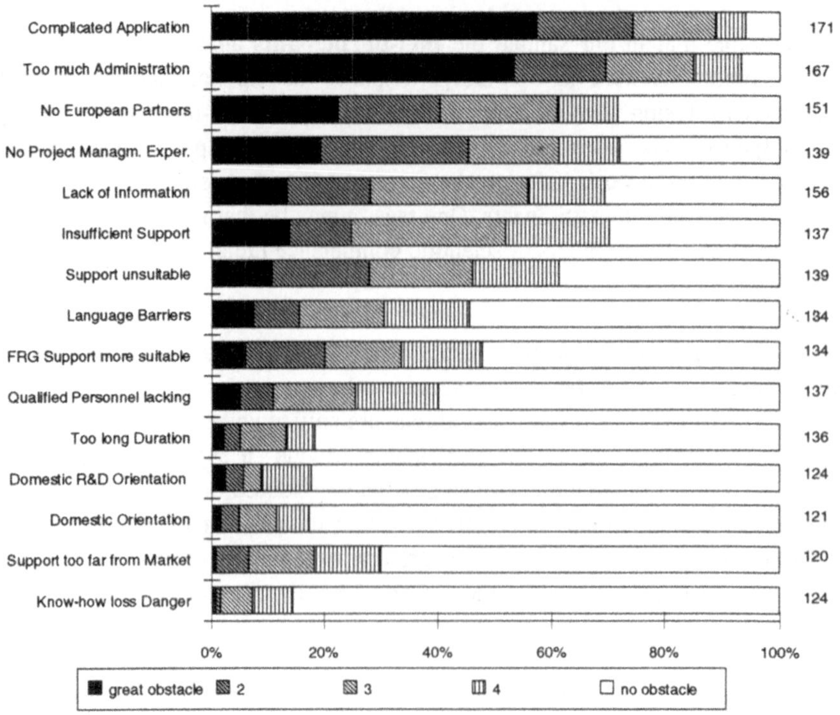

Nevertheless the **great majority** of non-participants in HEIs (87%) replying to the questionnaire would like to **take part in EC programmes.** There is also a small minority of non-participants (ca. 12%), however, that only wish to participate in German national RTD support.

Participants in EC projects, who have already gained some experience, also complain about certain problems and barriers in applying and in carrying out projects, but only to a very small extent. Thus it may be assumed that the obstacles they name (application;

administrative complexity) can also be overcome by non-participants, if **two main problems** can be removed: (1) **rejection of proposals** due to lack of EC funding and (2) an **insufficient international orientation** of the HEIs.

5.7 Future Development of National and Community Policy

The majority of HEIs in the survey consider the **extension of German RTD support** to be very important (65% of 150 non-participants and 67% of 170 participants). There are

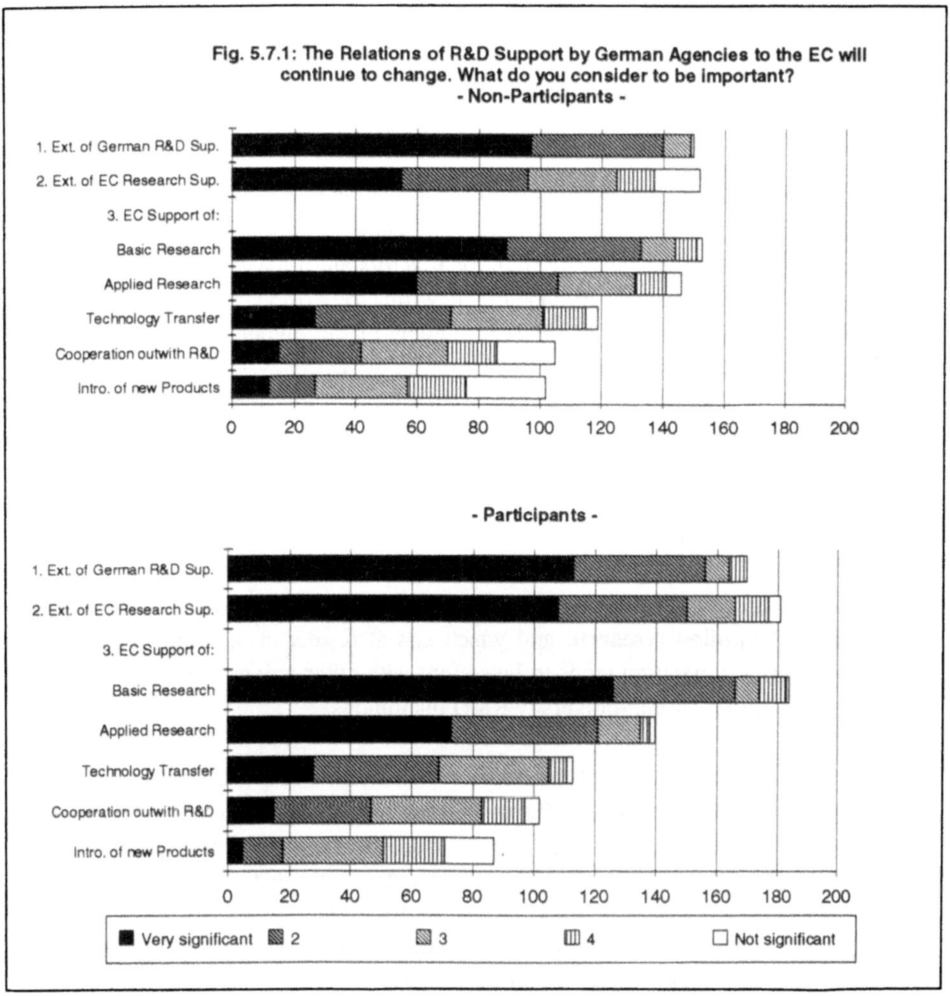

more differences of opinion regarding the strengthening of EC RTD support as a whole. **Participants** in the higher education sector consider the extension of **EC RTD support** to be more important than non-participants, who attach less significance to it. Since the main emphasis in universities is on **basic research**, the interest of the HEIs questioned is clearly centred on basic research in EC projects and understandably tends to decline as the market orientation of EC support increases (cf. Figure 5.7.1).

The development most hoped for by HEIs interested in EC project participation is a substantial **increase in the support budget** of the Framework Programme (especially in basic research activities), in order to improve the **acceptance rates** for proposals in future.

5.8 Summary and Conclusions

German university research is relatively well developed and traditionally well funded. In the last 20 years the importance of "third party funding" for research financing has continually increased, partly because the traditional actors and financers of research policy (in particular the DFG) have been joined by new financers (the BMFT, industrial enterprises), and partly because the public basic funding of the higher education sector has now been stagnating for years. In this situation, EC RTD policy has gained importance since the mid 1980s, slowly at first and then more rapidly, but until now it continues to play only an auxiliary role.

Strategic Role of EC Projects

The participation of German HEIs is concentrated in EC projects which are concerned with **basic and applied research**, and which aim at results in the **long term**. In the project consortia, cooperation tends to be sought with other universities, and to a lesser extent with enterprises or non-university R&D institutions.

EC projects are an integral part of the research portfolios of HEIs; the topics are mainly in the **core area of their research fields** and the projects serve to extend existing research fields. In projects, the research groups and institutes supply their existing expertise, but also to some extent - particularly in the science-oriented programmes - start to work in areas that are new to them; in this respect the EC support can be said to have initiated additional, original research activities. Interviews indicate that the generation of new scientific knowledge occurs more in projects oriented towards basic research, which are predominantly financed from national research funding; on the other hand, in EC projects - particularly in the programme lines "I&C Technologies" and "Modernisation of Industry" - there is a dominant orientation towards applications, due apart from anything else to their purpose of encouraging partnerships with industry.

The EC funding is financially necessary for the participating HEIs in order to carry out their research; many projects would **not have been possible without EC support.** The financial significance of the EC support will increase in the future, as the German HEIs, against a background of feared financial cutbacks in Germany, are pinning their hopes on the EC as a compensatory financial source.

Reasons for Participation

The **financing of research work** is the central reason for participation in EC projects. HEIs also consider it important that EC RTD programmes **complement the topics** of German support programmes and regard them as very important for **cooperation with European universities**. HEIs participate in EC projects with the main objectives for participation of enhancing their scientific skills and reputation.

Although EC participation will become more attractive for German HEIs in future, they fear that the projects will involve a great deal of complex administration, and that the time-consuming process of submitting project proposals will be a strain on their resources.

Impacts on Higher Education Institutes

It is the scientific impacts of EC RTD support that are mentioned most frequently by HEIs; they consider that participation in projects definitely increases their **scientific capabilities**. Secondly, their **willingness to cooperate** with European universities has increased. They have succeeded in establishing new research fields, and their contract research has increased.

In the respect of supporting European cooperation in research, the RTD programmes of the EC fulfill an important function for German HEIs. It is obvious that a great deal of importance is attached to inter-university cooperations in particular.

Non-participants in the Higher Education Sector

The great majority of non-participants in the survey are planning to participate in EC RTD programmes within the next five years, and thus represent **a large potential**; it is an open question whether the funding necessary for this will be available in the future. Approximately one quarter of non-participants questioned have already applied for EC support but were not successful.

On the one hand, the behaviour of non-participating HEIs provides an explanation for their non-participation in EC projects: they have **less of a European orientation** than participants, and their R&D activities are more strongly **directed towards experimental development**; however, since in EC projects universities are expected to perform research and supply fundamentally oriented specialist knowledge, this orientation to development would seem to limit their participation chances. On the other hand, the time-consuming and difficult process of submitting a proposal, and the administrative and organisational complexity of projects, are regarded by HEIs as a serious obstacle to participation.

Future Development of European RTD Policy

In future, the majority of HEIs would like to see in the first place an extension of German RTD support, and in the second place a strengthening of EC RTD support. The main change that HEIs would like to see in the Framework Programme is **more support for basic research.**

6. The Importance of EC RTD Policy and its Influence on Non-university R&D Institutes

The following chapter is concerned with R&D institutions outside the higher education sector. This is a heterogeneous sector which has been enlarged and has undergone some changes as a result of German unification. In the new Länder (formerly the GDR), a research landscape which was previously quite differently structured has been being assimilated since 1990 into the western research landscape of the Federal Republic of Germany. Non-university research institutes, referred to in the following as **non-university R&D institutes (RDIs)**, are made up of (cf BMFT 1993):

- 16 **national research centres** (Großforschungseinrichtungen/GFEs) in the old Länder, and 8 branch institutes and 3 new national research centres in the new Länder, accounting for a total of ca. 5% of German expenditure on R&D; these GFEs perform basic and applied research and have big items of equipment designed for special fundamental high technology research; these institutes receive varying rates of public support which may be up to 90% of basic financing; 90% of this is contributed by the federal government and 10% by the Länder in which the institutes are located.

- The **Max Planck Society** (Max-Planck-Gesellschaft/MPG), which currently has 62 research institutes in the old Länder, and in the new Länder 2 institutes, 2 branch institutes and 28 working groups on a temporary 5-year basis; the MPG conducts mainly basic research in selected areas of the natural sciences and the arts, and accounts for just under 2% of German R&D expenditure; the MPG receives institutional support from the federal government and the Länder; the level of this basic financing in 1991 was 83%.

- The **Fraunhofer Society** (Fraunhofer-Gesellschaft/FhG), which at present has 39 institutes in the old Länder and 8 institutes, 1 sub-institute and 12 working groups in the new Länder; the FhG performs mainly applied research for industry and for public institutions, and accounts for just under 1% of German expenditure on R&D; the FhG institutes receive support only as a limited, success-dependent proportion of their basic financing (under 30% in 1991), and have to cover their costs by contract research and innovation-oriented services.

- The "**Blue List" institutes** (Blaue-Liste-Einrichtungen/BLE): 48 **joint institutes of the federal and Länder governments** in the old Länder and 34 in the new Länder, accounting altogether for just under 4% of German R&D expenditure; the Blue List institutes are very heterogeneous regarding their tasks, financing, size, locations and legal status, and are financed on a fifty-fifty basis by the federal government and the "Land" in which they are located. The Blue List institutes are an important instrument of federal research policy, as through them the host Länder can build up specialised research centres of supraregional importance.

- 54 **Federal and Länder research institutions with research tasks** in the old Länder, and 3 institutions and 30 branch offices in the new Länder; these have the status of public institutions and, among other official duties, perform research for the state. In some of these institutions the proportion of R&D is less than 10%, whereas others were founded expressly for research and testing purposes; generally these institutions are financed fully and institutionally by the federal government; third party financing is an exception, but some federal institutions with R&D tasks do participate in EC projects.

- 68 **institutes of the Confederation of Industrial Research Associations** (Arbeitsgemeinschaft industrieller Forschungseinrichtungen/AiF), receiving a share of ca. 1% of German R&D expenditure; these institutes are all under the umbrella of the AiF, to which at present 102 industrial research organisations belong; the AiF institutes perform sectorally oriented research and development close to industry, especially for SMEs.

- Numerous **private non-profit institutions close to** but outside **industry**, such as testing and consulting institutes, private R&D institutes and institutes associated with higher education institutes (HEIs).

Measured by their share of national R&D expenditure, these institutes taken together are equivalent in importance to the higher education sector: the 25% support share of non-industrial R&D in Germany is divided fairly equally between HEIs and RDIs.

This chapter presents the results of the written questionnaire and complementary personal interviews for the institutes. The response rate of institutes receiving the questionnaire was approximately 30% (176 replies received from a total of 586 participants in EC projects).

6.1 Characteristics of Participants and Non-participants in the Survey

Participants in EC RTD Programmes

The percentage distribution of the responding sample of 176 institutes with experience in EC projects is shown in Table 6.1.1. If this is compared with the shares of institutes in German research expenditure (see above), it becomes apparent that

- the representation of the national research centres (GFEs) and the FhG is roughly equivalent to their importance in the institute research landscape, and
- the MPG, the AiF and the Blue List institutes are under-represented - the MPG slightly, the AiF and the Blue List institutes seriously under-represented.

Table 6.1.1: Distribution of Responses from RDIs (N = 176)

	Number	%
GFE	49	27.8
MPG	19	10.8
FhG	27	15.3
AIF	4	2.3
BLE	6	3.4
Private	21	11.9
Others	46	26.1
No particulars	4	2.3

Together the large non-university research organisations in Germany (GFE, MPG, FhG, AiF) account for more than half the responses of institutes to the questionnaire. More than one third of replies were from a strongly heterogeneous mixture of private and other institutions including research-intensive engineering bureaux, testing and consulting institutes, private non-profit R&D institutes, institutes associated with HEIs and federal institutions with R&D duties. Since the tasks, sizes, financing and activities of the various non-university R&D institutions in the Federal Republic of Germany differ widely, this must always be borne in mind when interpreting the aggregated data; the **results only reflect trends and, due to the heterogeneity of the actors, may be very different for the various types of institutes**. For reasons of confidentiality, a more detailed disaggregation of the data from the questionnaires was not possible. In the written text (but without data) we have considered some organisations and R&D institutes more closely.

Table 6.1.2.: Size of the RDIs investigated

Academic Personnel (full-time employees)	Participants		Non-participants	
	absolute	%	absolute	%
Up to 10	16	9.3	10	18.5
11 - 49	52	30.2	24	44.4
50 - 149	46	26.7	18	33.3
150 and more	58	33.7	2	3.7
Total	172	100.0	54	100.0

More than 50% of responses are from institutes with more than 50 employees; of these, the majority are from institutes with more than 150 employees (cf. Table 6.1.2).

Approximately three quarters of institutes answering the questionnaire are from the following "**Regierungsbezirke**" (primary administrative divisions of the Länder): Köln and Düsseldorf (15.4%), Oberbayern (13.6%), Braunschweig (11.9%), Karlsruhe (10.8%), Berlin (10.2%), Darmstadt (6.8%) and Stuttgart (6.3%) (cf. Table A6.1.3 in the Annex). As well as HEIs (particularly technical universities such as München, Braunschweig, Karlsruhe, Stuttgart, Darmstadt), these areas are usually the sites of national research centres, MPG institutes, FhG institutes and other R&D institutes. The regional distribution of support funding thus reflects not so much the areas of high population density as the locational distribution of the technical and scientific research infrastructure in Germany.

The RDIs participate - dependent on their function - **very frequently in public support programmes** (EC participant sample more than 90%, non-participant sample more than 80%) (cf. Table 6.1.3). It is remarkable that when replying to the question on participation in other public programmes, participating RDIs most frequently name other public support programmes of the CEC (over 90%) - even more frequently than support of the Federal Ministry for Research and Technology (BMFT), which ranks second with over 80% (cf. Table 6.1.3). It is clear that in the area of project support EC support has now become very important.

Table 6.1.3: Is your Organisation Participating in (other) Publicly Supported RTD Programmes?

RTD Programmes	Participants		Non-participants	
	absolute	%	absolute	%
No	13	7.4	11	18.3
Yes, namely:	163	92.6	49	81.7
further EC projects	148	90.8		
Eureka projects	33	20.2	2	4.1
BMFT programmes	135	82.8	44	89.8
BMWi progammes	23	14.1	16	32.7
Länder programmes	49	30.1	28	57.1
other programmes	49	30.1	10	20.4

Multiple answers were possible.

Non-participants in EC RTD Programmes

The questionnaire for non-participants in EC programmes was answered by 60 RDIs. If non-participants are broken down according to (parent) organisations, 80% are found to be private institutions, Blue List institutes, AiF institutes and others. Only 20% of non-participants are FhG or MPG institutes or GFE laboratories. This structure is the reverse of the sample of participating institutes.

Over 50% of non-participants have less than 50 employees, and hardly any non-participating institutes have more than 150 employees. In this respect non-participants differ substantially from the group of participants (cf. Table 6.1.2), in which RDIs with more than 150 employees play a prominent role.

Both groups of RDIs make use of public RTD support programmes. Following BMFT programmes, which are ranked first (89.8%), the group of non-participating institutes gives preference to programmes of the Länder governments (57.1%) and the BMWi (the Federal Ministry for Economic Affairs) (32.7%), neither of which are so important for the group of EC participants (cf. Table 6.1.3). There are regional concentrations of non-participants in the Köln/Düsseldorf area (ca. 18%), Dresden (10%), the "Land" of Berlin (ca. 7%), Leipzig (ca. 7%), Freiburg and Tübingen (5% each). Compared to the sample of EC participants, regions in the new Länder and regions without technology- and science-oriented higher education institutes tend to be represented in the non-participant sample.

All in all, the sample reflects the structure of non-participating institutes: as most GFEs, Max Planck institutes and Fraunhofer institutes participate in EC programmes, only the remaining institutes in this group of actors could be addressed and could reply to the questionnaire as non-participants.

6.2 Importance and Strategic Significance for Participants

Having defined the group of RDIs questioned, this chapter continues with a description of the importance attributed to EC support by the RDIs with experience of EC participations. First the projects given by participants as background experience in their answers will be characterised (programme affiliation, financing, constellation of partners, position within the project of the answering institute, time horizon, type of research). Then interrelationships will be presented that have been observed between the R&D portfolios, technological competences and R&D financing of the RDIs on the one hand, and the EC support projects on the other.

6.2.1 Characterisation of EC Projects

The **distribution of projects over various EC programmes** can be seen in Table 6.2.1.1. It is noticeable that there are many more answers relating to the BRITE/EURAM programme than to ESPRIT projects. Going by the data of the basic population, one would have expected this to be the other way round. Otherwise, the distribution of projects mentioned by answering institutes adequately represents the EC data on the distribution of EC projects over EC programmes.

The EC generally finances half the project costs. About 90% of institutes stated that they used their **own funds** to finance the share of project costs not covered by the EC funding. In some individual cases, residual funding is received by GFEs, Max Planck institutes and Blue List institutes especially from the federal and Länder governments, and by the Fraunhofer institutes from industrial partners.

Table 6.2.1.1: Structure of the answering R&D Institutes in EC Programmes

EC Programmes (II. Framework Programme) 1987-1991)	Survey Sample (n=175)		Basic Population (n=556)	
	absolute	in %	absolute	in %
1. Quality of Life	**36**	**20.6%**	**92**	**16.5%**
e.g.: Human Gen. Analysis	1	0.6%	1	0.2%
Medicine/Health	2	1.1%	8	1.4%
Radiation Protection	14	8.0%	42	7.6%
STEP/EPOCH	19	10.9%	41	7.4%
2. Information and Communication Technologies	**37**	**21.1%**	**179**	**32.2%**
e.g. AIM	4	2.3%	11	2.0%
DELTA	1	0.6%	5	0.9%
DRIVE	2	1.1%	12	2.2%
EURET	0	0.0%	0	0.0%
ESPRIT II	22	12.6%	126	22.7%
RACE I	8	4.6%	25	4.5%
3. Modernisation of Industries	**39**	**22.3%**	**75**	**13.5%**
e.g.: B.C.R	2	1.1%	0	0.0%
BRITE/EURAM	31	17.7%	63	11.3%
Raw Materials	6	3.4%	12	2.2%
4. Biological Resources	**11**	**6.3%**	**18**	**3.2%**
e.g.: Agriculture	1	0.6%	0	0.0%
BIO/BRIDGE	9	5.1%	16	2.9%
ECLAIR	0	0.0%	1	0.2%
FLAIR	1	0.6%	1	0.2%
5. Energy	**18**	**10.3%**	**71**	**12.8%**
e.g.: Decommissioning	2	1.1%	6	1.1%
Fusion	1	0.6%	0	0.0%
JOULE	11	6.3%	46	8.3%
Radioactive Waste	4	2.3%	14	2.5%
TELEMAN	0	0.0%	5	0.9%
6. Science and Technology for Development	**3**	**1.7%**	**18**	**3.2%**
7. Marine Resources	**3**	**1.7%**	**10**	**1.8%**
e.g.: FAR	0	0.0%	0	0.0%
MAST	3	1.7%	10	1.8%
8. Improvement of S/T Cooperation in Europe	**21**	**12.0%**	**93**	**16.7%**
e.g.: DOSES	0	0.0%	0	0.0%
EUROTRA II	1	0.6%	9	1.6%
Large-scale facilities	2	1.1%	1	0.2%
MONITOR	2	1.1%	7	1.3%
SCIENCE	16	9.1%	71	12.8%
SPES	0	0.0%	1	0.2%
VALUE	0	0.0%	4	0.7%
9. Other (related) Programmes	**7**	**4.0%**	**0**	**0.0%**
Total	**175**	**100.0%**	**556**	**100.0%**

The institutes have an average of 6.7 project partners (cf. Table 6.2.1.2), with numbers varying from 1 to 10 per project and in exceptional cases going up to 80 project participants. The partners are most often higher education institutes, especially in other European countries. However, European research institutions and enterprises, and partners within Germany (enterprises, HEIs and RDIs) are also mentioned.

Table 6.2.1.2: How many Partners have participated in this EC Project?

	RDIs	
	M	se
Firms from Germany	0.7	1.9
Firms from EC countries	1.3	2.2
HEIs from Germany	0.3	0.7
HEIs from EC countries	1.7	3.3
RDIs from Germany	0.4	0.6
RDIs from EC countries	1.3	2.0
Other organisations from Germany	0.1	0.4
Other organisations from EC countries	0.3	1.1
Total	6.7	8.1

M = Mean; se = standard error

Just under half the RDIs (48.3%) state that they are main contractors in the projects, 43.7% are contractors; approximately 8% are associates and subcontractors (cf. Table 6.2.1.3).

Table 6.2.1.3: Your Position in the Project is that of ...?

	RDIs		all Participants	
	absolute	%	absolute	%
Main contractor	84	48.3	225	38.8
Partner	76	43.7	305	52.6
Associate	12	6.9	36	6.2
Subcontractor	2	1.1	14	2.4
Total	174	100.0	580	100.0

The main emphasis in the institutes' EC projects is on **applied research** (56.6%), but 23% of participating RDIs designate their projects as focusing on **basic research** (cf. Table 6.2.1.4). In ca. 20% of answers experimental developments and demonstration projects are named as the main project topics. Individual analysis by sub-groups of RDIs

- based, however, on relatively low case figures - emphasize the wide variety of tasks and activities:

- Among **contract research institutes**, especially the FhG and private R&D institutions, application-oriented research dominates, with complementary activities in other R&D areas.

- **AiF institutes** are engaged in equal proportions of applied research and experimental development.

- The work of the **"Blue List" institutes** is one third in applied research, one third in basic research and one third in experimental development.

- As well as application-oriented research, **GFE laboratories** also carry out substantial activities in basic research.

- **Max Planck institutes** mainly carry out basic research.

Table 6.2.1.4: Which Type Characterizes Best Your Share in the Whole Project ?

	RDIs	
	absolute	%
Basic research	41	23.4
Applied research	99	56.6
Experimental development	22	12.6
Demonstrators and diffusion	13	7.4
Total	175	100.0

6.2.2 Strategic Role of EC Projects

Importance for RTD Activities

More than two thirds of RDIs state that the EC project is a **definite part of their R&D portfolio**. Only one third of the RDIs see little (23.8%) or no (8.1%) connection with their other projects (cf. Table 6.2.2.1). Projects with a **medium- or long-term**

orientation predominate (over 80%) over projects with a short-term orientation (15.6%).

RDIs usually place the EC project topics in the **strategic core area of their own technological competence** (40.9%). An important part (27.4%) is played by projects in research fields of possible great future importance for the participating institute (potentials, core areas). However, projects in areas of technology not closely connected with the core areas of institutes' technological competence, mentioned in 30% of answers, also play a considerable part.

Table 6.2.2.1: Composition of the R&D-Portfolio

What Connection does the EC Project have to other R&D Projects in your Organisation?	RDIs	
	absolute	%
Integrated part of R&D portfolio	119	69.2
Few connections to other R&D projects	40	23.3
No connections to other R&D projects	13	7.6
Total	172	100.0
Time span of this Project		
Short-term	23	15.6
Medium-term	65	44.2
Long-term	59	40.1
Total	147	100.0
What Role does the EC Project play for your technological Competence (technical standing)?		
Strategic core area	67	40.9
Potential new core area	45	27.4
Just one technology among others	49	29.9
Rather peripheral technology area	3	1.8
Total	164	100.0
What Role does the EC Project play in your Strategy?		
Establish new research areas	29	19.0
Extension of a research area	94	61.4
No significance	16	10.5
Diversification	12	7.8
Cost reduction	2	1.3
Total	153	100.0

For nearly two-thirds of RDIs replying to the questionnaire (61.4%), the EC project serves to **extend a business or research field**. One fifth (19%) are participating in EC projects in order to build up a new field (cf. Table 6.2.2.1).

It emerges from the interviews that RDIs generally have a broad technological spectrum which is not rigorously compartmentalised or static. Their acquisition efforts are very often aimed at expanding their technological competence. Internationally known institutes try to influence EC programme development on the basis of their own technological competence, and at the same time orient the development of their own competence towards the programme definitions. Since project acquisitions often take place decentrally, the **EC projects can often only be assigned ex-post to the core area or the marginal activities of institutes.**

Financial Importance

The role played by EC projects in financing the R&D budgets of the RDIs is shown in Table 6.2.2.2[31]. Just under half the budget on average is covered by the institute's own resources (including non-project-related public institutional funding/basic financing). In addition, one quarter comes from national support programmes (federal 21.1%, Länder 7.1%). EC funding constitutes ca. 11% and rose, in the period from 1989 to 1991, from 8.3% to 10.9%, whereas over the same period of time the average share of institutes' own resources in their total annual budget went down from 48.7% to 46.5%.

RDIs **expect the upward trend in EC support to continue in future** (cf. Figure 6.2.2.1): in the next 5 years they expect the importance of the EC to increase most, relative to other clients, followed by enterprises in Germany. Whereas EC support funds are expected to increase, expectations about support from German federal sources are divergent: decreases, increases and a constant level are all forecasted. The same is true for support by the Länder. This reflects the inherent tension, also apparent in the interviews, of a situation in which - despite the constraints placed on public budgets by the economic recession and German unification - RDIs feel that more emphasis should be placed on the support of research and technological development.

31 The data in the table must be interpreted with caution, as the financing of the RDIs is extremely diverse. Whereas Max Planck institutes receive about 90% of the basic financing, the average institutional support received by the Fraunhofer institutes is 30%; moreover, the relative importance of contracts from enterprises is considerably higher for contract research institutes.

Table 6.2.2.2: What Share in the Financing of your R&D Budgets do the following have

	Mean 1991 in %	in future decreasing		in future constant		in future increasing		Total	
		abs.	%	abs.	%	abs.	%	abs.	%
Own funds	46.5	45	42.1	52	48.6	10	9.3	107	100.0
Own contracts	0.3	1	6.3	14	87.5	1	6.3	16	100.0
Contracts from firms	8.7	3	5.1	16	27.1	40	67.8	59	100.0
Contracts from HEIs/RDIs	0.8	2	10.0	17	85.0	1	5.0	20	100.0
EC Funding	10.9	7	5.9	48	40.7	63	53.4	118	100.0
Federal Republic funding	21.1	39	37.5	43	41.3	22	21.2	104	100.0
Länder funding	7.1	18	27.3	33	50.0	15	22.7	66	100.0
Other sources	4.0	5	8.1	25	40.3	32	51.6	62	100.0
Total	100.0								

Fig 6.2.2.1: Which Institutions or Agencies will become more important for the Execution of your RTD plans in the next 5 years ?

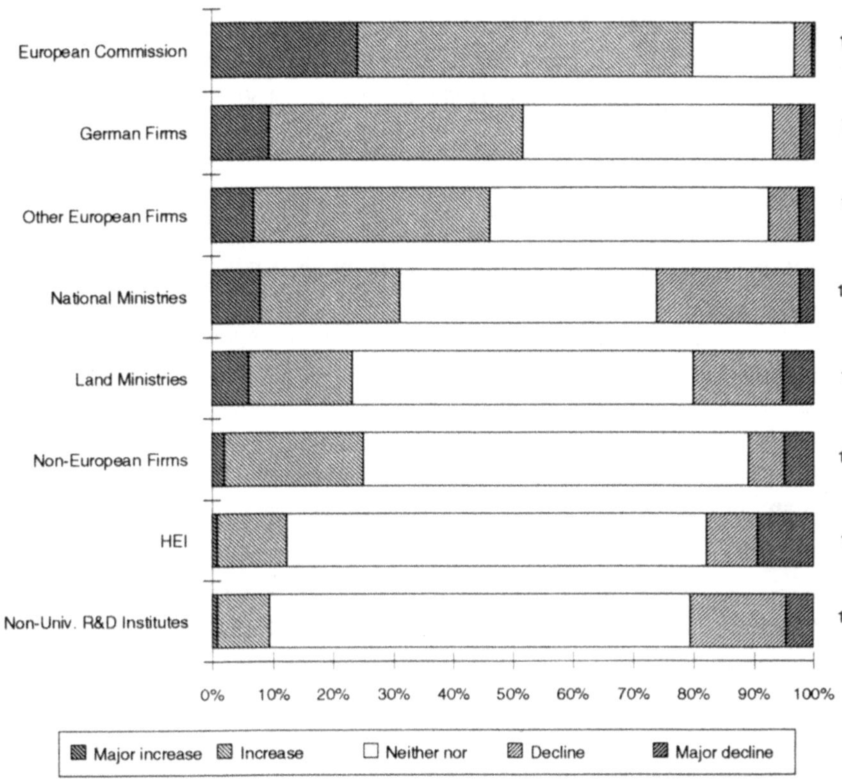

However, there are great divergences between the RDIs in the financing of their R&D budget: **Fraunhofer institutes** answering the questionnaire cover only ca. 20% of their budget from their own resources, as opposed to just under 70% for Max Planck institutes and more than 50% for GFEs. Federal project funding received by Fraunhofer institutes has decreased from ca. 30% (1989) to ca. 26% (1991) whereas EC funding received has increased from ca. 7% (1989) to ca. 8% (1991). For **Max Planck institutes**, federal and EC project funding are of approximately equal importance, at 12-13% each (1989 and 1991). Federal project funding of **GFEs**, whose own resources have remained constant, has gone down from ca. 26% (1989) to 24%, and EC funding has risen from ca.5% (1989) to ca. 7% (1991). Enterprises as contract clients have a higher relative importance for contract research institutes than for institutions oriented towards basic research.

The data relating to structure of R&D budgets are based on answers from the individual institutes replying to the questionnaire, which have usually participated in several EC projects; the average share of EC funding for all institutes of an organisation is certainly lower. In a few cases, in interview statements, large industrially-oriented contract research institutes advanced EC funding into first place as a source of financing.

General Reasons for Participation

The most important reason for carrying out R&D projects with EC funding is their use as a **further source of financing** (78.9%) (cf. Table 6.2.2.3). It also plays an important role that the **topics of the Community programmes** are regarded as **complementing the German support programmes** (66.9%), and that within the technology fields of the projects **European cooperation** (61.7%) is considered important. Another argument for participation in EC projects given by one third of the RDIs is that EC projects enable them to give their R&D a longer-term orientation. In some cases EC projects are used to build up cooperative business relations within Europe outside the R&D projects.

Table 6.2.2.3: Why are you carrying out R&D Projects with EC Funding?

	RDIs	
	absolute	%
Further funding source	138	78.9
Enables long-term orientation	59	33.7
Necessary supplement to German support	117	66.9
European cooperation is important	108	61.7
German programmes unsuitable	14	8.0
Exploration of new research areas	10	5.7
Establish European cooperations	36	20.6
Other reasons	13	7.4
Total	175	100.0

Multiple answers were possible.

The strategic **financial importance of EC RTD support** for RDIs investigated is assessed as high: more than half the institutes (55.1%) would not have carried out the project at all without EC funding (cf. Table 6.2.2.4). Projects were able to be carried out in extended form (33.5%) and in a small number of cases faster (14.8%). In 14.2% of answers the project would have been carried out without funding, but with different goals. Only ca. 6% of RDIs state that the project would have been carried out regardless of EC funding. This indicates that the so called "**free rider effect**" is small according to the self-appraisal of the RDIs answering.

Table 6.2.2.4: Which Role did the EC Funding Play ?

	RDIs	
	absolute	**%**
Independent of funding	11	6.3
Faster through funding	26	14.8
Larger scale possible through funding	59	33.5
Other goals possible through funding	25	14.2
Not at all without funding	97	55.1
Total	176	100.0

Multiple answers were possible.

In two points, answers contain **potential contradictions** requiring closer examination:

- firstly, EC support is seen as a necessary complement to national programmes, although the German programmes are described as appropriate,

- secondly, more than half the projects would not be carried out without EC funding, although in answers they are placed predominantly in the strategic core area of the institutes' competence.

In the interviews it emerged that EC programmes complement national support in the sense that

- **to some extent** they **compensate cuts in national support** (this refers only to complementation in the financial sense, and not to content);

- EC support is **closer to industry** compared with support by the BMFT (especially in information technology);

- participation in EC projects **contributes to the formation of a "European technology community"**; it is considered essential to belong to this community (which is, however, referred to by many as a "closed shop") in order to acquire further contracts and to influence future EC research topics;

- moreover, RDIs with a high degree of institutional basic financing do not necessarily require additional external funding in order to carry out their strategic R&D projects; **EC support makes possible additional R&D activities** in important fields in which the institute is already working with its basic financing.

In the area of institutes financed by contract research, there is a danger that, without EC funding, projects may not be carried out despite their strategic importance. Contract research institutes are dependent on third party financing, which is essential in order for them to perform R&D projects.

Frame Conditions of the EC

In interviews with contract research institutes it was stated that although EC RTD support is an important complement to national actions as regards programmes, financing and the support of cooperation, the **frame conditions are not appropriate for contract research institutions** such as e.g. the Fraunhofer Society:

- The EC support model on a **cost sharing basis makes access** to EC partipation **much more difficult for Fraunhofer institutes**; the low rate of institutional basic financing (ca. 30%) is neither sufficient, nor was it originally intended, for participation in EC projects on a cost sharing basis, and is not consistent with the EC model of "cost shared actions".

- Contract research institutes without a high degree of basic financing are thus **unable to participate more often in EC projects** so long as the "cost shared actions" model continues to stipulate 50% own financing by participants. This affects primarily the FhG, whose EC funding expanded to 22 million DM per year by 1991, and has been stagnating at this level since 1992.

- At present the "residual financing" is only covered in some institutes by direct contributions from industry. There is thus a danger that the emphasis in R&D will **shift even closer to applications and to the market**; projects which have a higher risk factor because they are oriented towards the future, and projects connected with preventive research are hardly ever financed jointly by industry, and cannot so often be engaged in by contract research institutes. The problems of residual financing cannot be covered by additional financing from industry since this appears rarely and leads to a shift of the research areas.

The **Max Planck Society and the GFEs consider that the EC Framework Programmes are not sufficiently suited to their own R&D programmes**. They recognise that in accordance with the subsidiarity principle EC RTD support has tasks that are complementary to national support, especially regarding new knowledge-oriented basic research; however, on the grounds of **decentralisation and quality assurance** they suggest that

- scientific quality should be taken as the outstanding criterion for approval of R&D projects,
- new, autonomous support organisations should be set up at European level, committed to the principles of scientifically oriented research, to judge applications for support in the area of new knowledge-oriented basic research, and
- in the area of application-oriented R&D, suggestions for the selection of assessors should regularly be obtained from autonomous science organisations in the Member Countries.

6.3 Reasons and Aims for Participation, Results and Success Factors

The RDIs see EC projects as increasingly important for their research activities and for their integration into a future "European technology community". The following section presents the aims envisaged and achieved in the projects, the relationship of these aims to the support aims of the EC, and the factors mentioned as contributing to the success of projects.

The answers of the RDIs were grouped into three EC programme lines (corresponding to the programme lines of the Second Framework Programme) and one combined group, and were evaluated in these groups (cf. also Chapter 6.2.1):

- The programme line **"Information and Communication Technologies (I&C Technologies)"** includes over 20% of answers to the questionnaire, with the emphasis clearly on ESPRIT and to a lesser extent RACE.

- The programme line "**Modernisation of Industry**" also includes over 20% of answers, mainly relating to BRITE/EURAM.

- The programme line "**Improvement of European Science and Technology Cooperation**" (**S&T Cooperation**) accounts for just over 10% of answers, with the main emphasis on SCIENCE; a separate evaluation is necessary because of the specific orientation of this progamme line.

- The remaining answers relate to several different programmes (particularly STEP/EPOCH, RADIATION PROTECTION, JOULE), but because of the small numbers involved were not evaluated separately, but combined in an "**other areas**" group.

Project Aims

Aims of projects in the programme line "**I&C Technologies**" are primarily increasing the knowledge base, further participation in other EC projects, enhanced skills of R&D personnel, enhancing scientific reputation, and better cooperation with European HEIs/RDIs (cf. Figure 6.3.1). Developing new methods, designing prototypes, cooperative relationships with enterprises and innovative development of products are considered as aims of medium importance.

For the programme line "**Modernisation of Industry**" the order of importance is similar, but cooperation, both with HEIs/RDIs and with enterprises, is placed higher (cf. Figure 6.3.1), whereas enhanced skills and further EC participations, are less important. In both programme lines the aims mentioned by the institutes express the orientation of programmes such as ESPRIT, RACE, BRITE/EURAM towards **applied research and development, and towards industry**.

By contrast, in the programme line "**S&T Cooperation**" purely scientific aims such as improving the science base, enhancing scientific reputation, higher qualification and preparing PhD theses play a greater role (cf. Figure 6.3.1). Cooperation with HEIs/RDIs is ranked second (together with scientific reputation), while cooperation with enterprises has very little importance. Nor are product and process innovations, new or improved services or prototypes being aimed at. The results reflect the orientation of this programme towards better cooperations between European scientific institutions and towards scientific aims.

Fig. 6.3.1: Which of the following Goals did you aim for in this Project? (R&D Institutes)

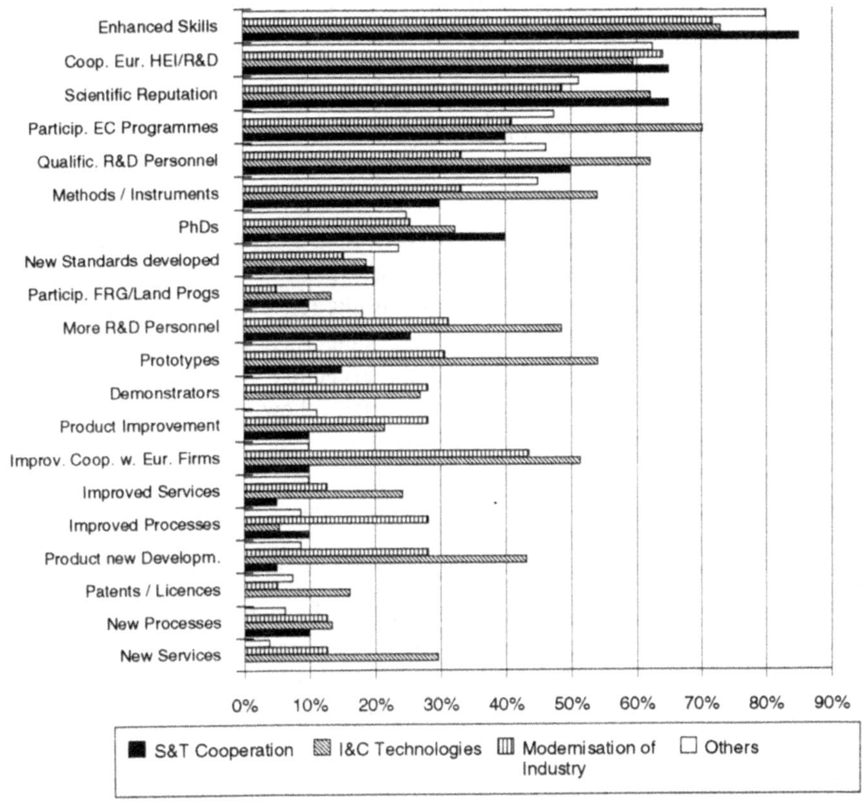

The diverse aims of the programme lines in the Second Framework Programme are reflected in the different rankings in answers from the RDIs. Taking into account the various aims of the programmes, it can be stated that for the RDIs as a whole **increasing the knowledge base** (by improved knowledge, qualification of R&D personnel and new methods), together with **technological and scientific cooperation with higher education institutes/R&D institutes** in Europe, occupies first place. Aims also very frequently mentioned are **enhancing scientific reputation** and **further participation in EC programmes**.

Project Results

The RDIs fully succeed in attaining their aims, with the ranking of aims being approximately reflected in the results achieved (cf. Figures 6.3.2 to 6.3.5). The **attainment of scientific results** such as a more extensive knowledge base, scientific reputation, qualification of personnel and **better cooperation with European HEIs/RDIs** head the list. The strengthening of cooperation with enterprises was more often aimed at, and more often achieved, in the industrially oriented EC programmes than in the programme line "S&T Cooperation", in which directly exploitable technological results play no part (cf. Figure 6.3.2).

The **medium to low importance of aims** such as new or improved products or process technologies, development of new standards or services in the industrially-oriented programme lines is reflected in the attainment of results (cf. Figures 6.3.3 to 6.3.6). The **time horizon** should also be noted here: whereas the scientific aims have usually been attained on completion of the project (with the exception of PhD theses), RDIs only expect technical results within the three year period following project completion. Increased knowledge, better relations with HEIs and scientific reputation are built up directly in the work process. New products and standards, on the other hand, can only be expected at the end of a process in which the relevant EC project represents only the initial stage.

It is nevertheless striking that directly exploitable technological aims (products, standards, demonstrators) are seldom put on a comparable footing to scientific aims and/or results. From some interviews the explanation emerges that many EC projects in which RDIs participate have a much more distant time horizon with respect to conversion of results into marketable products and processes (5 years or more, rather than up to 3 years). These long-term projects often involve preliminary development for the investigation of alternative options (electro versus combustion engine, digital versus analogue technology, etc.) and at this stage product development is still very much in the background. The aim of such projects is thus to present the technological possibilities.

This explanation was reinforced in one interview with a large contract research institute by the statement that it is mainly in long-term process innovations (automation, process organisation) that institutes are brought in, and that for reasons of secrecy developments for future products in industrially-oriented EC programmes are only carried out in purely industrial consortia, and with the perspective of strategic alliances.

Fig. 6.3.2: Which Results did you achieve in this Project? (S&T Cooperation, RDIs)

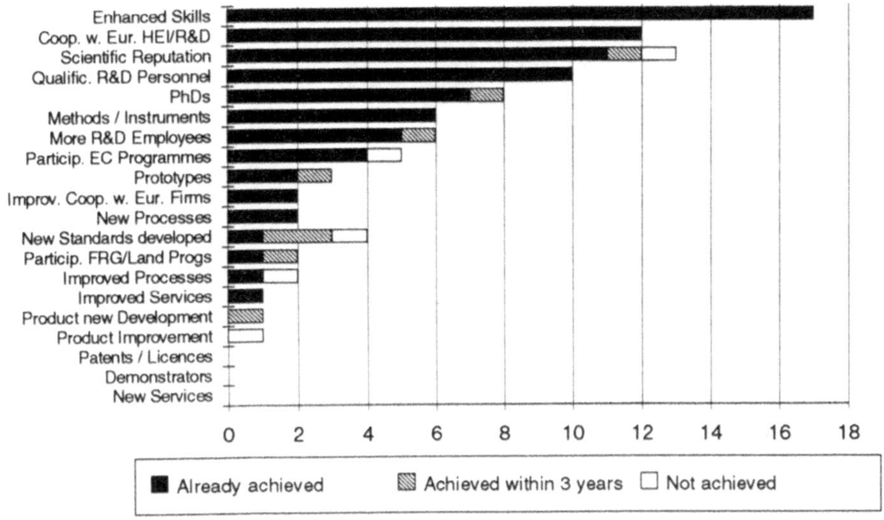

Fig. 6.3.3 Which Results did you achieve in this Project? (I&C Technologies,RDIs)

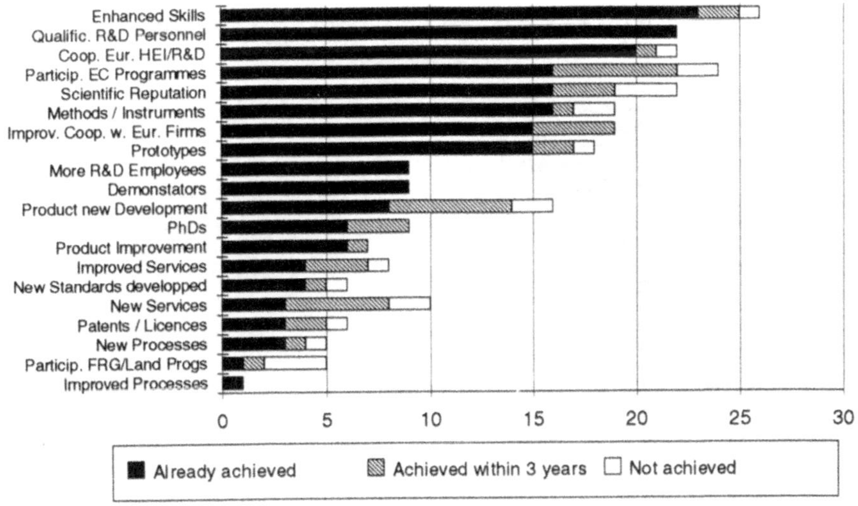

Fig. 6.3.4: Which Results did you achieve in this Project (Modernisation of Industry; RDIs)

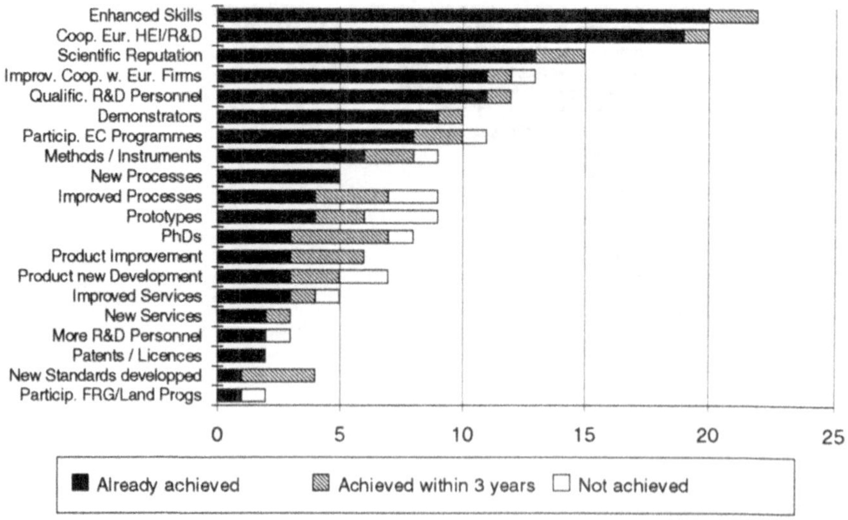

Fig. 6.3.5: Which Results did you achieve in this Project (Others; RDIs)

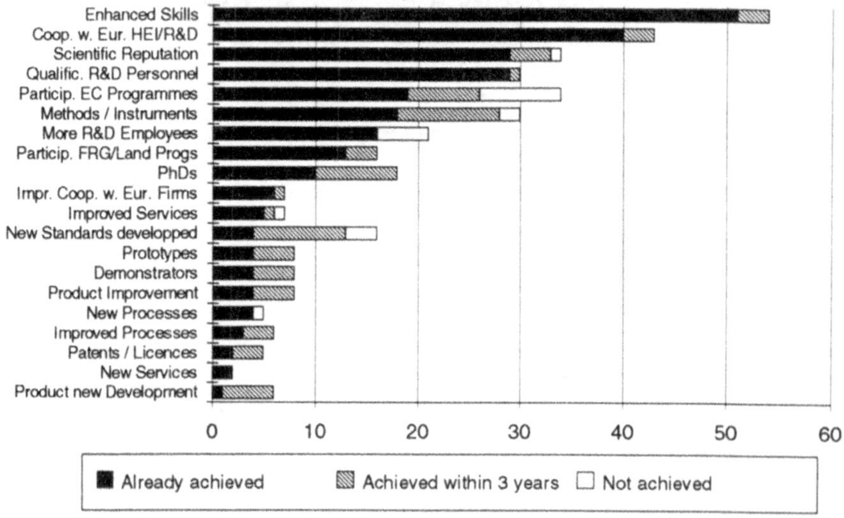

All in all, over 97% of the RDIs state that they have achieved the initial scientific and technological goals; more than three quarters give an achievement level of 80%. The attainment of scientific aims is manifested in an average of ca. one PhD thesis or one contribution to a book produced per project, and 6 to 8 articles in journals, conference papers and working papers per project. This output places the RDIs ahead of the higher education sector in terms of documentary results.

A **differentiated analysis of the institute groups** yields results generally similar to the results for the institutes as a whole, with the differences that

- **contract research institutes** mention technical goals such as the innovative development of products more often than the other groups,

- **GFE laboratories** mention relatively often (34%) the aim of increasing their number of scientific personnel. Interviews would seem to indicate that this is due to the situation of the GFE laboratories, which are being asked by the BMFT to cut down on staff; EC funding could compensate or moderate the effects of cuts in basic financing.

Figure 6.3.6 gives an overview of the **most important success determinants for aim achievement**. These include

- the technological competence of the partners,
- agreement on the project content in the consortium,
- the adequate defining of technological goals,
- the participating institute's own project management and the project management of the main contractor, and
- the composition of the consortium.

Changes in the institute's own priorities and aims in the course of the project were assessed mainly as a positive factor for success, but similar changes made by the partners tended to be assessed as **inhibiting** the success of projects. Partners not sticking to time schedules, the communication aspect (time, difficulty and expense) and re-organisation by partners were also described as obstacles to project success (cf. Figure 6.3.6). Different cultures and mentalities within the consortium were considered to favour rather than impede project success. Project monitoring by the EC was mainly assessed as positive for the success of EC projects.

Fig.: 6.3.6: Which Factors have helped or hampered the Success of this EC Project or led to scrapping of Project?

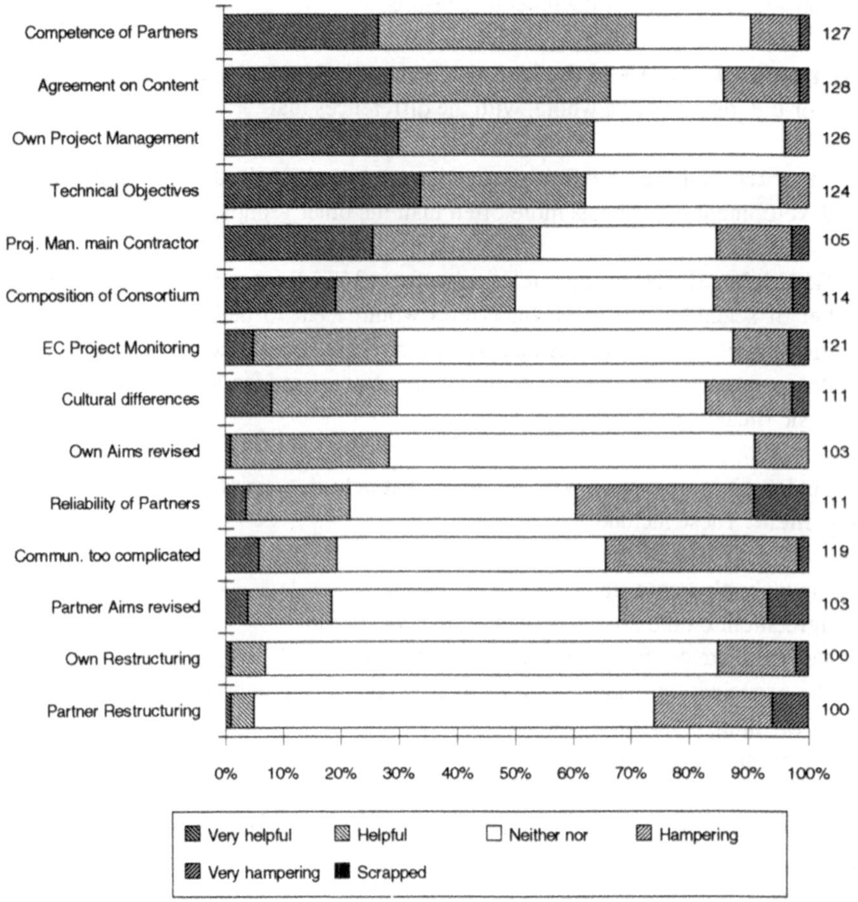

To sum up, the RDIs definitely give strong priority to scientific aims and cooperations, and **less weight to market-oriented technological** aims. Technological aims are pursued with a longer-term time horizon. The aims of RDIs in participating in EC projects are predominantly fufilled. However when considering this result, the great heterogeneity of the RDIs should be borne in mind: Max Planck institutes and the GFE laboratories have their own research focuses, which diverge to some extent from the EC programmes, and are oriented towards fundamental research, whereas the emphasis of the contract research institutes coincides to a greater extent with the orientation of the EC programmes.

Particularly the priorities of the science-oriented institutes **correspond only partially to the aim of the RTD support of the European Community**, since the institutes clearly give higher priority to scientific aims than to directly transferable technological aims. The EC's aim of improving European cooperation is fulfilled more within the science system than with industry.

6.4 Impacts on Participating R&D Institutes

6.4.1 Economic, Technical and Scientific Impacts

The RDIs regard **increasing their scientific capability**, **establishing new research fields** and **increases in contract research** as the most important impacts of their participation in EC projects. Other significant effects, in order of importance, are **supplying new services** (some institutes, for instance, offer project management services based on their experience in EC projects), **increasing the R&D workforce**, better **recruiting of scientists** from the EC and **changes in technology strategy**.

These seven very important or important impacts of EC projects also come to the fore if the answers of institutes are differentiated according to programme affiliation (three programme line groups and the "other areas" group), but there are variations in the order of importance (cf. Figure 6.4.1.1 to 6.4.1.4). The greatest differences are observed in the **programme line "S&T Cooperation"**, in which more importance is given to the impacts of recruiting scientists from the EC and increasing the number of R&D personnel (cf. Figure 6.4.1.1). This is due to the orientation of the "S&T Cooperation" programme line, and the **SCIENCE** programme[32] contained in it, towards stimulating cooperation and the exchange of European research scientists.

32 More than 75% of the institutes' answers in the programme line "S&T Cooperation" relate to the SCIENCE programme.

Fig. 6.4.1.1: Which Impacts has this EC Project had on your Organisation? (S&T Cooperation; R&D Institutes)

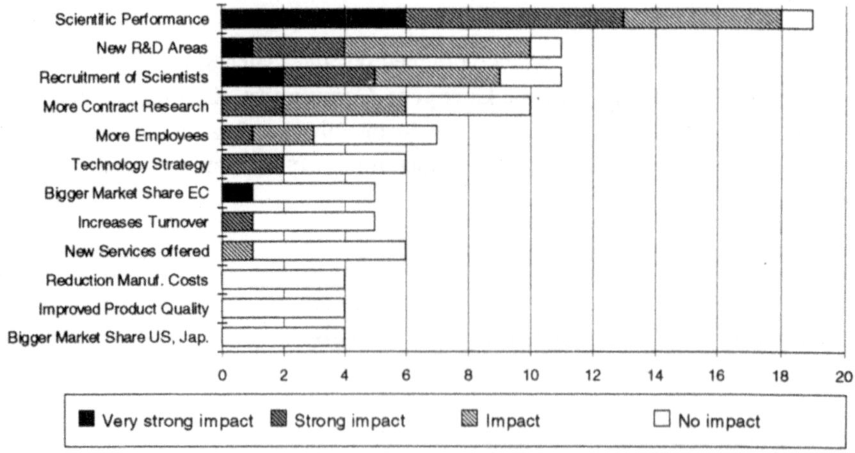

Fig. 6.4.1.2: Which Impacts has this EC Project had on your Organisation? (I&C Technologies; R&D Institutes)

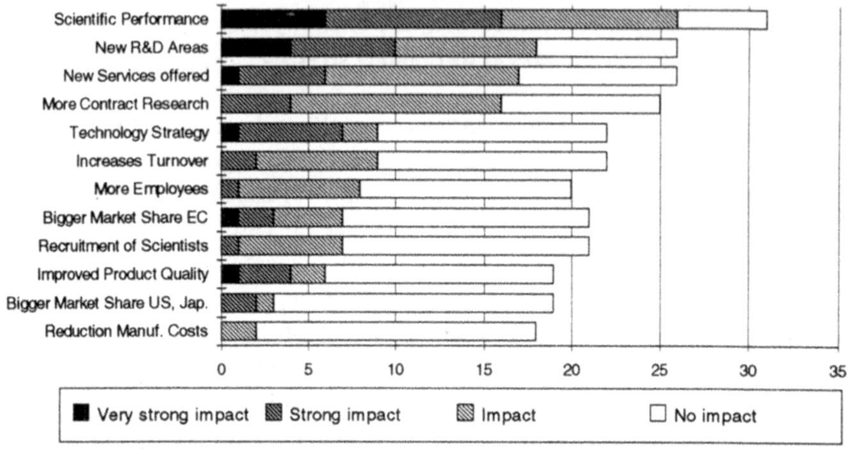

Fig. 6.4.1.3: Which Impacts has this EC Project had on your Organisation? (Modernisation of Industry; R&D Institutes)

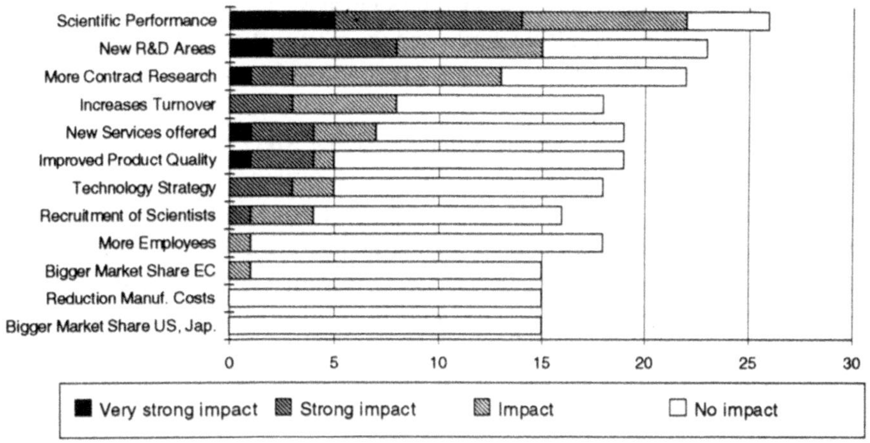

Fig. 6.4.1.4: Which Impacts has this EC Project had on your Organisation? (Others; R&D Institutes)

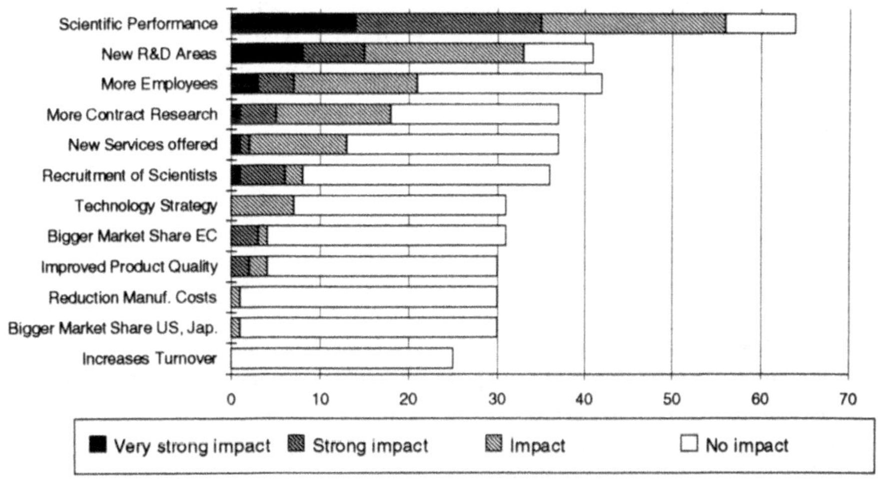

Increases in turnover were mentioned by a few institutes in the programme lines "**I&C Technologies**" and "**Modernisation of Industry**" as the only fairly relevant economic impact. The few answers in the "**economic impacts**" category (cf. Figure 6.4.1.2) were from private R&D institutes and from institutes engaged in contract research. For the Max Planck institutes, the GFE laboratories and the Blue List institutes, these economic variables are not relevant, due to their high degree of basic financing.

All in all, **scientific impacts** (scientific capability, more R&D staff, exchange of scientists) and **strategic impacts** (new research fields, new services, more contract research) dominate the institutes' answers. Economic impacts relating to turnover, market share and thus to competitiveness are only of secondary importance. This result is corroborated by very emphatic statements in interviews **rejecting a direct use of research work for competitive economic success.** Research and development generally, and in EC projects in particular, is only seen as **one** of many prerequisites for long-term competitive economic success in industry. Innovative behaviour patterns of enterprises and economic and legislative frame conditions are regarded as considerably more important for competitiveness. The correlation formulated in the Second Framework Programme between promoting science and technology and increasing the long-term competitiveness of European industry was treated with reservation in discussions with the science-oriented institutes.

The interviews also made it clear that the RDIs would not wish to do without the RTD support of the Community. The benefits derived from **European cooperation and securing a position in the developing "European technology community"** are considered as far outweighing the frictions involved in applying for support and working in a consortium.

6.4.2 Impacts on the Cooperative Behaviour

In most cases, RDIs participating in projects supported by the EC have already cooperated with other project partners (just under 95%). Cooperation is **primarily with** HEIs and other RDIs **enterprises** as clients occupy a **secondary** place (cf. Table 6.4.2.1). Cooperation with competitors and consultants is mentioned least often. The same general response pattern emerges in a comparison of cooperation behaviour relative to Germany, the EC, the USA, Japan and the rest of the world.[33]

33 However, the number of responses decreases: 143 answers mention cooperations in Germany, as many as 145 name Europe, but cooperations outside Europe are only mentioned in 88 answers, 94 name cooperations with partners in the USA and 48 mention Japanese partners.

Table 6.4.2.1: Cooperations with other Organisations in RTD (Participants)						
	Germany	EC	Rest of Europe	USA	Japan	Rest of World
Suppliers	17	10	4	8	4	1
Customers	32	29	11	14	3	5
Competitors	11	11	3	1	2	1
Universities	125	115	64	69	30	39
RDIs	101	112	62	59	27	34
Advisors	25	8	2	2	0	
Network	34	44	17	15	9	11

Partnerships in EC projects usually originate from already **existing contacts** (cf. Figure 6.4.2.1). **Conferences and publications,** and **cooperation in other EC projects**, are important sources in the search for partners. Mediation by other partners is slightly less important. By contrast EC mediation, EC data banks, public liaison offices and EC offices of scientific organisations appear to play a relatively minor role in the search for partners.

This result is hardly surprising, since **direct personal contact is usually the decisive factor in the search for partners**. Impersonal media have only a secondary importance. However, these services provided by the EC and German federal advisory offices are used more by first-time applicants to the EC than by applicants with previous EC experience.

Table 6.4.2.2: What Impacts did your Cooperation with European Partners have?		
	absolute	%
No further cooperations	6	3.4
Decreased willingness to cooperate	12	6.8
Increased willingness to cooperate	117	66.5
Extension of R&D cooperation	117	66.5
Extension of cooperation	19	10.8
Total	176	100.0

Multiple answers were possible.

For the great majority of institutes, previous cooperation in EC projects has led to a **general increase in willingness to cooperate and/or to extend R&D cooperation** (66.5% in each case; cf. Table 6.4.2.2). R&D cooperations have most often been extended to include **other EC projects** (over 90% of answers) but also to include research cooperations in other types of publicly supported R&D projects and in joint research laboratories. Cooperations extending beyond the field of R&D (10.8%) include some individual instances of cooperations in product development, production, marketing and sales. No willingness (3.4%) or less willingness to cooperate (6.8%) following participation in an EC project was mentioned only in exceptional cases.

Fig. 6.4.2.1: What Significance do the following have when looking for Partners?

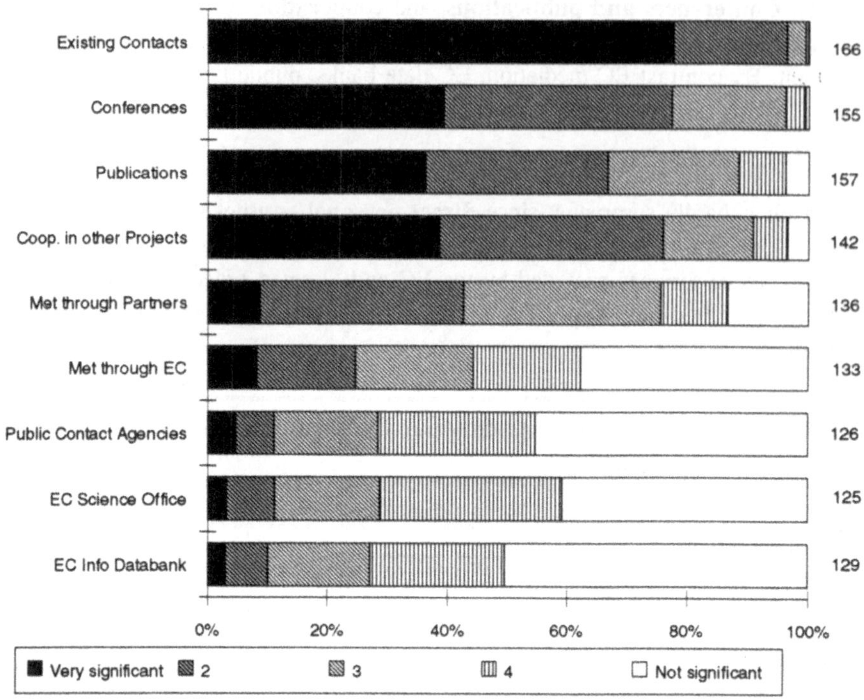

The results of the questionnaire on the impacts on the cooperative behaviour of the RDIs correspond to the results of the interviews, which emphasize the positive impacts upon institutes' readiness to engage in European cooperations as the greatest advantage of European science support compared to national actions. Cooperation at a scientific level is considered very **important in preparing the way** for a **"European technology community"** and, in the long term, improving Europe's competitive position.

6.4.3 Social and Environmentally Relevant Impacts

To complement the analysis of the scientific, technical and economic impacts of EC support, institutes were also asked about generally positive social and environmentally relevant impacts of EC support.

Fig. 6.4.3.1: Has this Project had any further social or environmentally relevant Impacts?

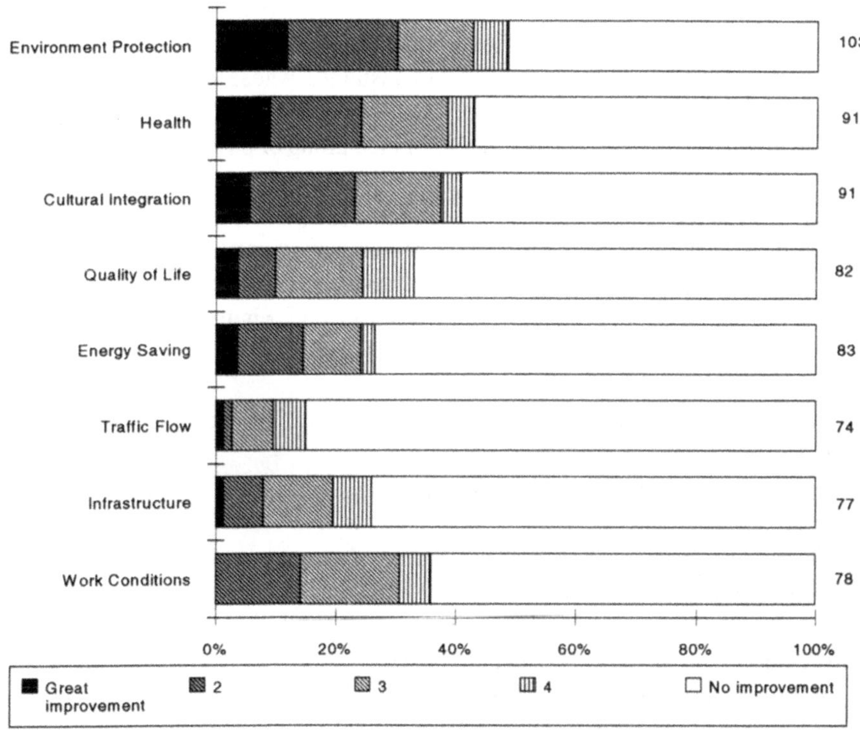

These general impacts **play only a limited role** in the answers of the non-university R&D institutes(cf. Figure 6.4.3.1). Although definite positive impacts on environment, health, cultural integration and energy conservation are mentioned, these tend to be attributable mainly to projects whose aims relate to these areas.

It emerges from the interviews that most project participants have no clear awareness of general impacts of this kind outside the project. General impacts are usually mentioned only if the **relevant support programme explicitly envisages these aims**. In programmes oriented mainly towards increasing industrial competitiveness, interviewees only give very vague answers to the question on general impacts.

6.5 Importance of EC RTD Policy for Non-participating R&D Institutes

In the following section, the results described above are contrasted with the innovative behaviour of institutes which have not yet received support from the EC (non-participants). Non-participants tend to be mainly from the groups of private R&D institutes, AiF institutes, Blue List institutes and other institutes whose tasks have a relatively strong national orientation. Size and cooperation behaviour are the other main aspects in which the group of non-participants differs from the group of participants.

In the questionnaire, 75% of RDIs which have not yet participated in EC programmes state that they have not applied. **Only 25% of non-participants in the questionnaire have already applied** to participate in EC programmes and been turned down. The following **barriers** were mentioned by non-participants (cf. Figure 6.5.1):

- the application process is too complex,
- the administrative and organisational aspects of EC projects are expected to be too expensive and time-consuming,
- financial support from the EC is too low.

As well as these main obstacles, the following factors were named as barriers of **secondary importance**:

- suitable European partners could not be found,
- deficits in information about the possibilities of EC support,
- no suitable EC programme existed for the projects,
- deficits in the international management experience of personnel.

On the other hand, possible obstacles such as the EC projects lasting too long on average, insufficient specialist capabilities among personnel, a predominantly inland orientation of R&D or the danger of "know-how drains" were **very rarely mentioned as barriers**.

Fig 6.5.1: Why does your Organisation carry out R&D Projects without EC Support?

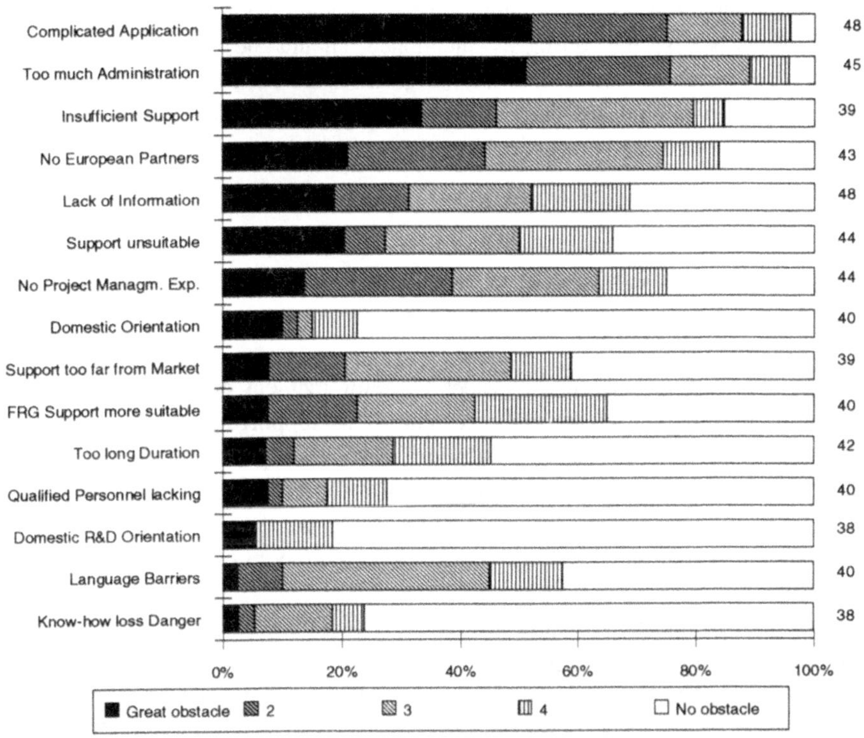

Taken in conjunction with the interviews, the answers seem to relate primarily to the fact that these **RDIs** are **comparatively smaller** in numbers of employees than the group of EC participants. This result also indicates that, despite its considerable rate of growth, EC support is **only making headway gradually as a source of financing** for the R&D projects of institutes, and that for first-time applicants in particular substantial barriers have to be overcome. For this reason EC support is often referred to as a **"closed shop"**.

Nevertheless, almost all non-participants questioned express the intention of **participating in EC programmes in future** (ca. 90%), mainly as contractors, associates or sub-contractors. However, just under half the institutes answering the questionnaire state that they would also be willing to act as main contractors in EC projects. This proportion seems very high, however, and would probably overtax the resources of some of the institutes.

Thus even among non-participants there is a predominantly positive attitude towards EC support. Nevertheless, the obstacles mentioned should be taken very seriously. Interviews, for instance, clearly describe the **great financial burden of applying** for support, which can amount to a third of the EC funds for an approved project; certainly, this is a development in the wrong direction.

6.6 The Future Development of National and Community Policy

The **non-participating institutes** consider the reinforcement of German RTD policy somewhat more important as the reinforcement of EC support (cf. Figure 6.6.1). Non-participants would like to see **EC support increased** primarily in the fields of **application-oriented research** and **fundamental research**. The support of technology transfer, market introduction and cooperations outside the EC (e.g. in the joint production of innovative products) are considered less important.

The **participating institutes** would also like to see an increase in both German and Community support; whereas **expansion of EC RTD support** is mentioned more often (cf. Figure 6.6.1). The growing importance of EC RTD foreseen by participants in the future is reflected in the answers to questions on the future financial role of EC funding: whereas German funding is expected to remain constant, the RDIs expect funding from the EC to increase. According to the institutes' subjective assessment of the future, a **proportional shift in emphasis is expected away from German support and towards EC support.**

Fig 6.6.1: The Relations of R&D Support by German Agencies to the EC will continue to change. What do you consider important?
- Non-Participants -

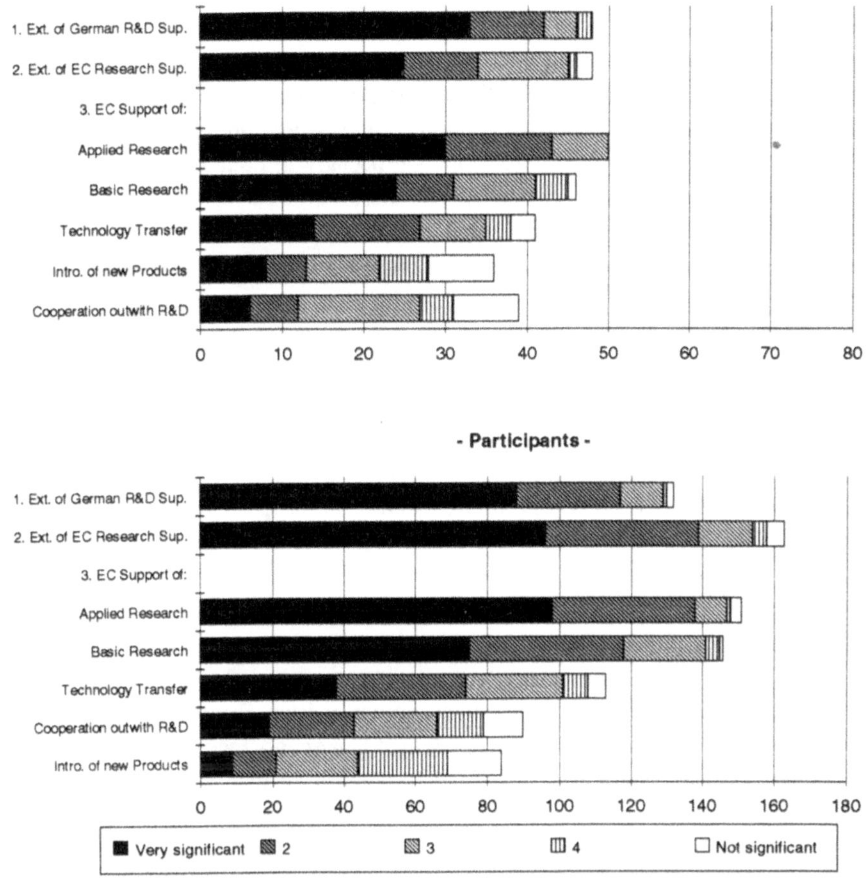

Participants would like to see a marked **strengthening of EC support** in the fields of **application-oriented research** and **fundamental research**. Evidently the present **orientation of the Second Framework Programme towards precompetitive research** (basic and applied research) corresponds very closely to the R&D activity of the participating institutes. The RDIs attach much less importance in this respect to technology transfer, cooperation outside R&D and the market introduction of new products.

6.7 Summary and Conclusions

Importance of EC Support for R&D Activities

In the majority of cases the EC projects are **definitely part of** the institute's **R&D portfolio**; they are situated primarily **in the strategic core area of technological competence** of the RDIs and secondly in potential new research areas. EC programmes are influencing the orientation of research topics especially in contract research institutes, but not so much in institutionally financed institutes.

Contract research institutes with a strong industrial orientation regard EC support as supporting their industrial networking (both with regard to projects and financing). Here EC projects support basic research for the long-term technology options of large companies, and also the product and process innovations, exploitable in the medium term, of medium-sized enterprises. **Institutes engaged in basic research** make use of EC support to strengthen their science and technology position and to extend their fields of research. All in all, in the general opinion of the RDIs questioned, the breakthrough of **EC support as a broadly accepted, transnational support action** in the field of science and technology took place in the course of the Second Framework Programme.

Reasons for Participation

The institutes take part in EC programmes primarily from **financial motives.** In the meantime, however, **belonging** to the growing **"European technology community"** plays an increasingly important role in order to keep up to date with both national and European research topics and scientific trends.

In 1991 the **share of EC funding in the research budgets** of non-university R&D institutes averaged between 1% and 10%; in the opinion of the RDIs, this can be expected to increase in the future. In the case of contract research institutes, these institutes are of the opinion that in view of the "cost shared action" principle of EC support their co-financing limits have been reached, and that this is preventing a further expansion of their participation in EC programmes.

Financial Shortages and Changing Frame Conditions

In the opinion of the institutes questioned, references to the expected growth of EC funding are used counterproductively in political discussions. The RDIs feel that **na-**

tional funding will be reduced in view of expected increases in EC funding, although the EC funding, despite its growth, will not compensate the national cuts.

With the increasing importance of EC support, non-university R&D institutes are becoming increasingly aware of restrictive EC frame conditions, which also affect national support. **Support quota restrictions** relating to **competition law** are criticised by contract research institutes. The **absence of an institutional support framework with autonomous science administration** for basic research and large research programmes (e.g. similar to the model of the DFG) is criticised by the Max Planck Society and the GFEs. Contract research institutes, on the other hand, appreciate the close link between basic and applied research in the EC projects.

Impacts of EC Projects

In the EC projects, the institutes are pursuing - and achieving - **primarily scientific** aims and **secondarily technical aims**. The institutes credit EC support with great success in **increasing willingness to cooperate** and **strengthening cooperation in R&D on the European level**.

There is less criticism by the RDIs of a lack of efficiency in EC R&D support. The procedures of programme definition, calls for projects, assessment and allocation as well as the running of the programme hardly differ any more from the national practices. **Programme definition and assessment of proposals** are considered to have **positive spin-offs** for the development of the "**European technology community**".

Important **points of criticism** made by the institutes refer to

- the **high refusal rate** of project applications and the ensuing wastage of economic resources,
- the disproportionality between the **lengthy** process of **programme definition** and the **short application times**,
- the **lack of coordination** that still exists **between national and European support regarding financing**, especially residual financing for institutes with no, or limited, sources of independent financing.

Taken as a whole, the institutes' assessment of EC research and technological development support is positive. The aims of EC support, to improve the science and technology base and promote European cooperation in research and technology, are considered to have been "**achieved**".

Need for Future Action

The need for future action seen by the **institutes** varies **according to their orientation and R&D activity**:

- **Contract research institutes**, because of their low level of institutional funding, have severe problems in providing their own financial contribution to projects on the "cost shared action" principle; the possibility of these costs being taken over by industrial partners in the EC projects causes an increasing orientation of contract research institutes towards market-oriented activities, while equally essential research-oriented activities are in danger of being neglected. There is a need for action by the CEC and by the Federal Government, if the participation of these institutes is not to be limited to EC projects that are exclusively application-oriented.

- **Institutes oriented towards fundamental research** advocate the institutionalisation of a scientific organisation at the European level, comprising on the one hand committees for the generating of science programmes and the approval of projects, and on the other policy advisory committees including various actors from the research landscape.

- The **individual procedures of EC support programmes should be more effectively designed**; unspecific calls for projects have led to floods of applications and high rejection rates; due to the high costs incurred by the application procedure, this constitutes a misapplication of the R&D resources of the institutes; moreover, the criteria for approval of R&D projects are considered to be obscure and impracticable.

- The fact that the Second Framework Programme is largely confined to **aims of competition policy** is considered to impose too narrow a limit upon the tasks of research, and to neglect, for instance, the challenges of preventive research.

In general, the non-university R&D institutes would like to see an extension of the Community RTD support and especially a strengthening in the fields of application-oriented and basic research.

7. RTD Support and the New Länder

Preliminary Remarks

When the previous German Democratic Republic (GDR) joined the Federal Republic of Germany in 1990, the European Community's Second Framework Programme was already running. Thus it was not to be expected that institutions in the new Länder would take part in this EC Programme to any great extent. Consequently, Chapter 7 can only give first indications of conditions that affect the use of EC RTD support which arise from the special situation in the new Länder.

For an understanding of these indications, a knowledge of the present situation in the new Länder is essential. Thus the analysis is preceded by a brief description of the special features of the research landscape there.

Only a very few participants in EC programmes replied to the written questionnaire; their answers will be evaluated briefly in a later section. This will be followed by a description of the importance of EC RTD support and barriers to participation from the viewpoint of non-participants in the new Länder.

7.1 The Special Situation of Research Institutions in the New Länder

Since 1989, the research landscape of the previous GDR has been fundamentally restructured and largely adapted to the FRG model (cf. Meyer-Krahmer 1992; ISI/IWH 1993). During this process there has been a radical reduction of R&D employees.

At the beginning of 1990, of ca. 132,000 R&D employees in the GDR, calculated according to the OECD method, ca. 74,000 were working in industry, 18,000 in the Academy of Sciences ("Akademie der Wissenschaften") and 14,000 in higher education institutes (HEIs) (see Table 7.1.1). In 1991 the total number of R&D employees was approximately 82,000 (cf. BMFT 1993, 610); following the closure of the Academy of Sciences and a dramatic drop in research activities in industry, the total today is probably considerably lower even than this.

Before the re-structuring of the system, most research in the science sector was performed by the Academy of Sciences, which had no teaching duties. The HEIs, on the other hand, played a comparably less prominent role in research. The institutes of the Academy covered all the tasks which, in the FRG, are performed by a differentiated research landscape situated between the universities and industry (like e.g. national research centres, Max Planck Society, Fraunhofer Society).

Table 7.1.1: R&D Personnel in the GDR Beginning of 1990 according to Areas (according to number of full-time equivalent - FTE)

	R&D Personnel			
	Total FTE	Researchers FTE	Technicians FTE	Others FTE
GDR	132341	57326	33887	41128
Industry	74052	29810	21241	23001
Construction	5527	3071	1209	1247
Transport	2738	1287	626	825
Post/telecommunications	1203	663	335	205
Agriculture, forestry, food	13321	5084	3458	4779
Trade/supplies	287	186	34	67
Health	1198	525	481	192
Academy of science	18285	7632	3364	7289
Universities	13986	8140	2656	3190
Other institutes	1744	928	483	333

Source: ISI/IWH 1993

The **Academy of Sciences** was dissolved following the unification of Germany (see Krull 1992). After intensive processes of scientific evaluation, only parts of the Academy personnel were transferred into the newly institutionalised model along FRG lines. In the **area covered by the BMFT** (Federal Ministry for Research and Technology) these are (cf. BMFT 1993, 23):

- 3 new **national research centres** and 8 branches of existing centres (for a description of these and other institutions in the FRG research landscape, see the beginning of Chapter 6),
- 24 new institutes and 4 branch institutes of **"Blue List" institutions** (BMFT area),
- 9 institutes and 12 working groups of the **Fraunhofer Society**,

- 2 **Max Planck institutes**, 28 working groups at universities, one organising body for the 7 arts centres of the Max Planck Society and one branch of a Max Planck institute.

Other Ministries[34] have set up research institutes, Blue List institutes and branches of existing institutes in the new Länder; additionally these have also founded their own research institutions. Altogether, Federal and "Länder" governments have created approximately 12,500 jobs in research outside industry (cf. BMFT 1993, 23).

The Fraunhofer Society has been particularly concerned to set up institutes at the interface between research and industry. The Max Planck Society has decided on the foundation or planning of up to 10 more institutes, aiming initially to set up, if possible, one Max Planck institute per "Land".

The reorganisation of **university research** is part of a process of fundamental restructuring of East German R&D. A restructuring programme for higher education and research in the new Länder requires funds amounting to 2.427 thousand million DM, to be provided by the Federal and Länder governments in the ratio 75 : 25 (cf. BMFT 1993, 24). The aims of the programme are a renewal of personnel and of research fields in the universities, the **strengthening of basic research** by incorporating previous Academy scientists and teams into the universities, or into research institutions jointly funded by the Federal and Länder governments, and securing the science infrastructure.

In the GDR, ca. 75% of R&D personnel were working in the industrial sector (by comparison: 68% in 1991 in the old Länder). To give a very rough description, **industrial R&D** in the GDR was organised in research centres by state industrial combines (sectorally organised state holdings) and mainly applied research was performed. These tasks were complemented by service functions resulting from the autarky efforts of the combines (development of supply components, etc.), which were often unconnected with research or had more to do with production planning and work preparation. In the struggle of enterprises for survival following the transition from a planned economy to a market economy, these research centres were mainly disbanded from their industrial combines and tried, as so-called "research companies" ("Forschungs-GmbHs"), to build up new future perspectives either as manufacturing firms in the high tech sector, as engineering and planning bureaux, as "re-integrated" central research departments in the

---

34 The Federal Ministry for Food, Agriculture and Forestry, the Federal Ministry for Economic Affairs, the Federal Ministry for Labour and Social Affairs, the Federal Ministry for Regional Planning, Building and Urban Development, the Federal Ministry of Education and Science, the Federal Ministry for Youth, Family, Women and Health, and the Federal Ministry for Environment, Nature Conservation and Reactor Safety.

enterprises that succeeded the combines, or as public, non-profit research institutes (cf. ISI/IWH 1993).

Fig. 7.1.1: Development of RTD Personnel in the industrial Sector of the New Federal States

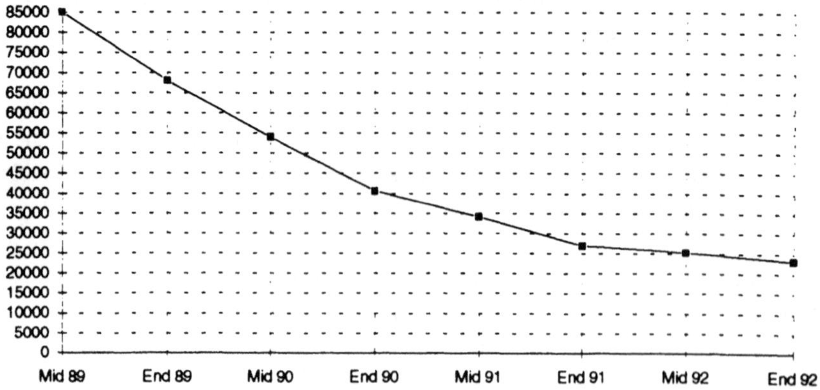

Full-time equivalent; until October 1990: GDR
Source: SV-Wissenschaftsstatistik, 1993

This **reduction of R&D personnel in industry** was particularly dramatic according to an investigation of the "SV-Wissenschaftsstatistik" (1993): Figure 7.1.1 shows the plunging figures from more than 80,000 in 1989 to less than 25,000 at the end of 1992. A study of the sectors of mechanical engineering, the chemical industry, electrotechnics and precision engineering/optics indicates that the decrease in these areas has been particularly drastic (cf. ISI/IWH 1993). The average shrinkage in these sectors is over 85% (cf. Table 7.1.2). This means that today only one seventh of the previous R&D workforce is still active in these sectors.

Table 7.1.2: Reduction of R&D Potential in Core Sectors

Sectors	R&D Employees (Persons)		Reduction from 1989 (100%) to
	1989	1992	
Chemical industry	9981	1616	16.2%
Mechanical engineering	8724	1950	22.3%
Electrical engineering	17752	1634	9.2%
Precision mechanics, optics, clock making	2380	366	15.4%

Source: ISI/IWH 1993

Both the Federal and the Länder governments agreed that the restructuring and building up process in the new Länder would need sustained supportive public action. Many **attempts have been made in economic and technology policy to promote the innovative capabilities of firms in the new Länder; these include:**

- building up an **infrastructure favourable to R&D**, including agencies for innovation support and technology transfer, innovation advisory offices for chambers of industry and commerce; centres for demonstration, consulting and technology transfer in new fields of technology;

- Initiation and strengthening of **technological competitiveness of firms in the new Länder**, including support of joint industrial research; easier access to BMFT project support under the specialised support programmes;

- Support of **technology-based firm foundations** and the initiation and expansion of technology and incubation centres;

- **Technologically oriented support of small and medium-sized firms**, including subsidising of R&D personnel costs and programmes for the support of innovation; support of contract research.

In 1992, approximately 607 million DM were made available by the BMFT and the BMWi for the securing and expansion of effective R&D capacities in industry in the new Länder (cf. BMFT 1993, 24). However, it is still not at all clear whether these funds will also be available in the medium term (3 to 5 years) as the basis for a self-supporting reconstruction of industrial R&D. In future, **the Länder** themselves will have to bear a **particular responsibility in this process** (cf. Kuhlmann/Berteit 1993).

At present the situation of many firms in the new Länder - despite public support - must be described as unsatisfactory: the **development of innovation strategies is still restricted** by prevailing conditions (incomplete privatisation processes; the disruption of regional and international supply and customer relations; the present cyclical decline). Frequently, R&D is still subordinated to short-term considerations, and concentrates on fast product developments which draw on the extensive fund of technological knowledge accumulated by the previous "combines". Imitations and improvements of traditional products generally predominate over innovations in research-intensive fields of technology (cf. ISI/IWH 1993).

With a view to the European Community and its RTD support, the task in the coming years must be to **direct the protagonists of the reduced and newly structured research landscape in the new Länder towards European partners and European**

cooperations, and make research competences in the new Länder productive for the long-term challenges of unified Germany and Europe. The following empirical results give indications of chances for the future and of existing barriers.

7.2 Empirical Basis of the Results

Responses to the questionnaire from **EC participants in the new Länder** (cf. Table 7.2.1) constitute only 2% of responses from all participants (9 out of 586). These answers were received from universities and research institutes but not from enterprises.[35] A statistical evaluation of answers from participants in the new Länder is therefore not feasible; it is only possible **to give some indications**.

Table 7.2.1: Response to Questionnaire

		New "Länder"		total	
		Number	%	Number	%
Participants		9	1.5	586	100.0
Non-participants		123	23.4	526	100.0
e.g.:	**HEIs**	51	41.5	189	35.9
	R&D institutes	21	17.1	60	11.4
	BIGs	13	10.6	54	10.3
	SMEs	36	29.3	217	41.3
	No particulars	2	1.6	6	1.1

The questionnaire addressed to **non-participants in the new Länder** elicited 123 responses, constituting ca. 25% of all responses from non-participants. Thus non-participants in the new Länder are adequately represented in the questionnaire. Nevertheless the actual numbers for the different groups involved, particularly in the case of the universities, are so small that a statistical evaluation of results differentiated according to groups is possible only to a limited extent.

35 The response for EC participants in the new Länder was received because, just to be on the safe side, a "participants" questionnaire was enclosed with the questionnaires addressed to non-participants; since only two participants from the new Länder were involved in the Second Framework Programme, the response is "accidental" (cf. Chapter 1).

The results of the written questionnaire are complemented by interview results from a recently completed study by the ISI (carried out jointly with the IWH, the Institute for Economic Research, Halle) on the situation in the new Länder (ISI/IWH 1993).

7.3 Indications from East German Participants

The fact that only nine answers were received from participants in EC RTD support actions confirms that until now EC support has not played a significant role for the new Länder. Some answers show that the partners from the new Länder were included in projects after they had started. The constellation of project partners was similar to other EC projects.

Replies relate to the Community support programmes BRITE/EURAM (3x), ESPRIT (2x), Environment (1x) and Technical Coal Research (ECSC) (3x). All nine participants who answered the questionnaire are also participating in support by the Federal Government and by the Länder. Almost all the projects would not have been carried out without EC support (the partner's own part). Answers about aims and results for participants in the new Länder show a distribution similar to the questionnaire in general. This applies also to answers on the search for partners and impacts.

All in all, up till the finishing of the survey for this study, the only recipients of EC project support in the new Länder have been research institutes which, **due to their existing relationships with partners in the West, were able to be integrated later in projects that were already running**. It is still too early for an analysis of the real importance of EC support for the new Länder. Interviews indicate that in the new Länder, even the "multipliers" such as the technology transfer offices of chambers of industry and commerce are only just building up their information about EC support programmes. Representatives of the new Länder ministries argue that in view of industry's present struggle for survival, **it takes too long for newcomers to overcome the obstacles to participation in EC support programmes**. This, they say, is a reason for concentration on the more quickly accessible national support schemes. The following section examines this view more closely from the viewpoint of non-participants.

7.4 Characteristics of Non-participants in the New Länder

R&D Characteristics

Almost all institutions replying to the questionnaire perform their own internal R&D (cf. Table 7.4.1). Internal R&D is mainly concerned with applied research (62% of responses). In addition, basic research on the one hand, and experimental development on the other, are each named by ca. one third of institutions as a main aspect of their R&D activities.

Table 7.4.1: Is your Organisation carrying out Research and Development Internally ?

Type of Organisation	HEIs (n=51)		RDIs (n=21)		BIGs (n=13)		SMEs (n=36)	
Yes, namely:	51	100.0%	21	100.0%	12	92.3%	31	86.1%.
Basic research	28	54.9%	8	38.1%	0	0.0%	1	3.2%
Applied research	39	76.5%	17	81.0%	7	58.3%	11	35.5%
Experimental development	13	25.5%	4	19.0%	9	75.0%	18	58.1%
Demonstrators and diffusion	3	5.9%	3	14.3%	3	25.0%	7	22.6%

Multiple answers were possible

In the higher education institutes (HEIs) and the non-university R&D institutes (RDIs), **applied research** dominates as the main aspect together with basic research. In enterprises **experimental development** takes first place. The fact that experimental development is mentioned relatively frequently as the main aspect of R&D activities agrees with information gained from the study on industrially-oriented research in the new Länder referred to above (ISI/IWH 1993). According to this study industrial R&D is concentrated on developing products suitable for the Western market by adapting old products to new market requirements. However, it should be pointed out that research in the former GDR was more practically-oriented before unification, and that the terms used in the questionnaire may have been differently understood.

R&D Funding

Three quarters of non-participants in the new Länder answering the questionnaire take part in **publicly funded RTD programmes** to finance parts of their R&D activities (cf. Table 7.4.2). By contrast, in the old Länder just under 90% of non-participants take part in regional or national support programmes. This result throws some light upon the situation in the new Länder: despite the extensive RTD support measures undertaken for the new Länder by the Federal Government, not all targeted institutions - especially SMEs - are being reached.

Table 7.4.2: Is your Organisation Participating in Publicly Supported RTD Programmes?

Type of Organisation	HEIs (n=49)		R&D Institutes (n=21)		BIGs (n=13)		SMEs (n=33)	
No	10	20.4 %	1	4.8 %	4	30.8 %	11	33.3 %
Yes. namely:	39	79.6 %	20	95.2 %	9	69.2 %	22	66.7 %
Eureka projects	1	2.6 %	0	0.0 %	0	0.0 %	0	0.0 %
BMFT support programmes	32	82.1 %	19	95.0 %	6	66.7 %	9	40.9 %
BMWi support programmes	6	15.4 %	8	40.0 %	3	33.3 %	11	50.0 %
Länder programmes	14	35.9 %	15	75.0 %	6	66.7 %	2	9.1 %
other programmes	19	48.7 %	4	20.0 %	1	11.1 %	6	27.3 %

Multiple answers were possible.

Correspondingly, the question on the relative shares of funding possibilities other than the institution's own resources or basic funding in the R&D budget was only fully answered by a part of the institutions; thus the data cannot be considered representative. With this reservation, the following financial structure emerges: the **R&D budget** in enterprises is financed mainly from their own funds (BIGs 85%, SMEs 74%). Higher education institutes finance more than 50% of their R&D budget from their own resources, but non-university institutes less than 20%. Support by the Federal Government plays the largest part for RDIs at over 40%, and is similarly important for HEIs, with a share of ca. 30%. Enterprises, on the other hand, finance only 10% of their R&D budget from federal support. Support from the "Land" governments plays a substantial role for HEIs and RDIs, making up 20% of their R&D budgets, but plays a negligible role for enterprises. As expected, EC support is not yet significant. 12 higher education institutes mention an average financing share of ca. 1% for EC support.

Taking into account the financing bottlenecks in the new Länder, this structure is already tending to align with the old Länder. The RDIs finance less of their R&D budget from their own resources or from basic funding than do those in the old Länder. However, a difference should be made here: institutions of the "recognised" R&D infrastructure, e.g. the Fraunhofer Society, are receiving a substantially higher level of basic funding in the new Länder during the initial building-up period than those in the old Länder. The **principal sufferers** from low levels of basic funding are the research capacities that have "fallen out" of the old GDR system and have not been completely absorbed into the new one (particularly the so-called "Forschungs-GmbHs"; cf. ISI/IWH 1993).

In future, **non-participants in the new Länder assume** that (see absolute data according to different R&D institutions Table 7.4.3)

- the share of their own resources will either decrease (47% of answers) or remain constant (40%) (only the RDIs expect an increase slightly more often),
- the share of EC support will go up (96%),
- the share of support from the Federal Government will grow (53%), as well as from the "Länder" governments (73%).

Table 7.4.3: Development of the Financing of the R&D Budgets in Non-participating R&D Institutions in the New Länder (Absolute Figures)

	HEIs (n=46)			R&D Institutes (n=19)			Enterprises (n=39)		
	⇙	⇨	⇗	⇙	⇨	⇗	⇙	⇨	⇗
Own funds	19	12	3	3	7	4	19	16	4
Own contracts	1	11	0	0	3	1	2	6	5
Contracts from firms	3	7	24	1	1	13	1	5	6
Contracts from HEIs	1	8	4	1	4	3	0	6	4
EC funding	0	0	28	0	0	12	0	2	11
Federal Republic funding	7	14	21	3	4	9	4	9	17
Länder funding	6	4	24	4	3	10	0	2	17
Other sources	1	11	9	0	4	3	1	5	5

⇙ decreasing
⇨ constant
⇗ increasing

Competitive Strategies

Considering the radical changes in the situation in the new Länder, answers to the question about competitive strategies are surprisingly "offensive" (cf. Figure 7.4.1; however, the low absolute figures of the answers should be borne in mind): approximately half the enterprises, in particular SMEs, are pursuing strategies such as technology leadership or concentration on market niches. The strategy of price leadership is mentioned infrequently. From interviews it emerges that in view of the struggle to survive in the short term, the innovation strategies of enterprises are often **limited to simply keeping in existence internal R&D capacities** that may ensure an innovative capability in the long term.

Fig. 7.4.1: Which competitive Strategies do you pursue in your Enterprise or Field of Business?

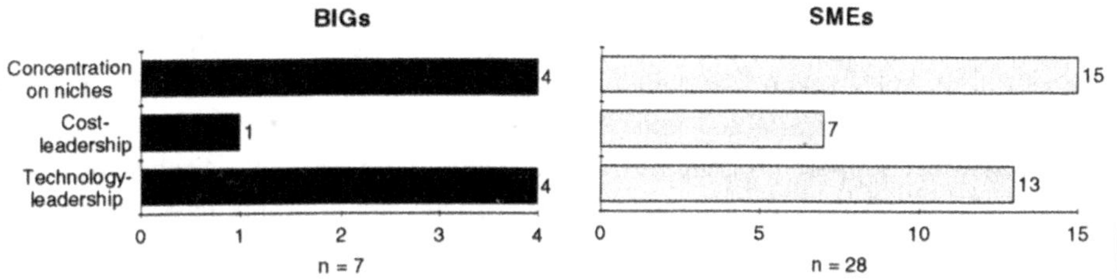

Cooperation Behaviour of Non-participants in the New Länder

Most non-participants in the new Länder are engaged in R&D cooperations with other organisations (cf. Table 7.4.4); only 15% state that they are not. Higher education and non-university R&D institutes tend to cooperate more often, and enterprises less often. Due to lack of responses it is not possible to determine whether transnational cooperations already play a role. In view of the obstacles mentioned, the deduction can be made that a general willingness to cooperate is already present, but that **the necessary partnership network is still lacking**. It also emerges from interviews that R&D performers in the new Länder attach great importance to RTD support generally, and to EC support in particular, in helping to remove this networking deficit

Table 7.4.4: Do you Cooperate in the Area of Research and Development with other Organisations (Contractual or Non-contractual R&D Cooperation)?

Type of Organisation	HEIs (n=51)		R&D Institutes (n=21)		BIGs (n=13)		SMEs (n=36)	
No	10	19.6%	2	9.5%	1	7.7%	6	16.7%
Yes, namely:	41	80.4%	19	90.5%	12	92.3%	30	83.3%
seldom	14	34.1%	7	36.8%	7	58.3%	24	80.0%
often	18	43.9%	11	57.9%	3	25.0%	5	16.7%
very often	8	19.5%	1	5.3%	2	16.7%	1	3.3%

Multiple answers were possible.

7.5 Future Importance of EC RTD Policy

Non-participants in the new Länder definitely represent a group of potential future participants in EC RTD support, although as a result of their situation their familiarity with Community programmes is still underdeveloped and they tend to give preference to more accessible support programmes of the Federal Government. Against this background, it is interesting to find out whether participation attempts have already been made, and if so what obstacles have cropped up, and to discover future intentions regarding EC participation.

17 of 123 non-participants in the new Länder (ca. 14%) have already applied for EC support, but without success. Figure 7.5.1 shows why non-participants carry out their R&D activities without EC support, and what barriers there are to participation in EC programmes. If the ranking of barriers by non-participants in the new Länder is compared with the ranking by non-participants in West Germany, there is general agreement about the **three most important obstacles**: the submission of proposals is considered too complex and time-consuming; so are the expected administrative and organisational aspects of the project management; there is a lack of information about EC support schemes.

More often than in the old Länder, **non-participants in the new Länder mention** their R&D staff's **lack of (international) project management experience**, the difficulty of finding **suitable European partners** and **language deficits** among their employees. Advisory organisations for the support of participation in EC programmes - such as the Euro-Info Centres - are thus particularly useful in the new Länder if they also make a point of offering possibilities for the improvement of languages and project management, and help in finding European partners, as well as information about support possibilities and applications.

All in all, the barriers mentioned by non-participants are not such as to cast any serious doubt upon the suitability of EC support for research institutions in the new Länder. This is confirmed by the fact that 92% of universities, 95% of RDIs, 73% of large enterprises and 72% of SMEs replying to the questionnaire would like to take part in EC projects.

When answering the question about areas in which RTD support at a national and European level should be extended in future (cf. Figure 7.5.2), **most importance** is attached to **extending German support.** The extension of Community RTD support is named less often. If the EC increases its funding, this should take place especially in the areas of **application-oriented research** and **basic research** from the view of the HEIs

Fig. 7.5.1 - 1: Why does your Organisation carry out R&D Projects without EC Support? (Non-Participants)
- East Part of Germany -

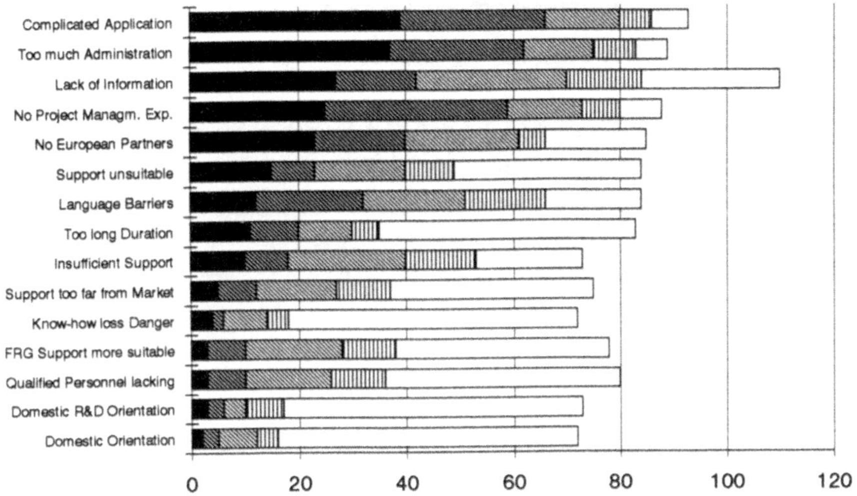

- West Part of Germany -

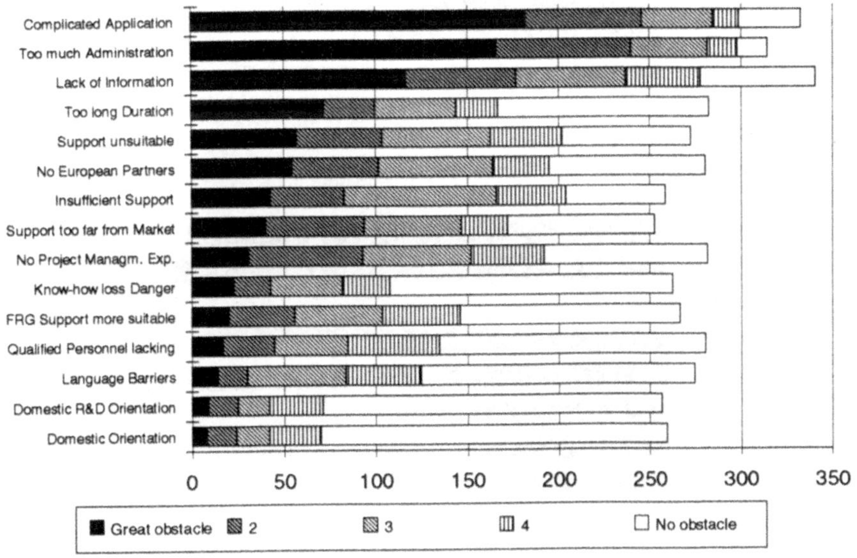

Fig. 7.5.2: The Relations of R&D Support by German Agencies to the EC will continue to change. What do you consider important?

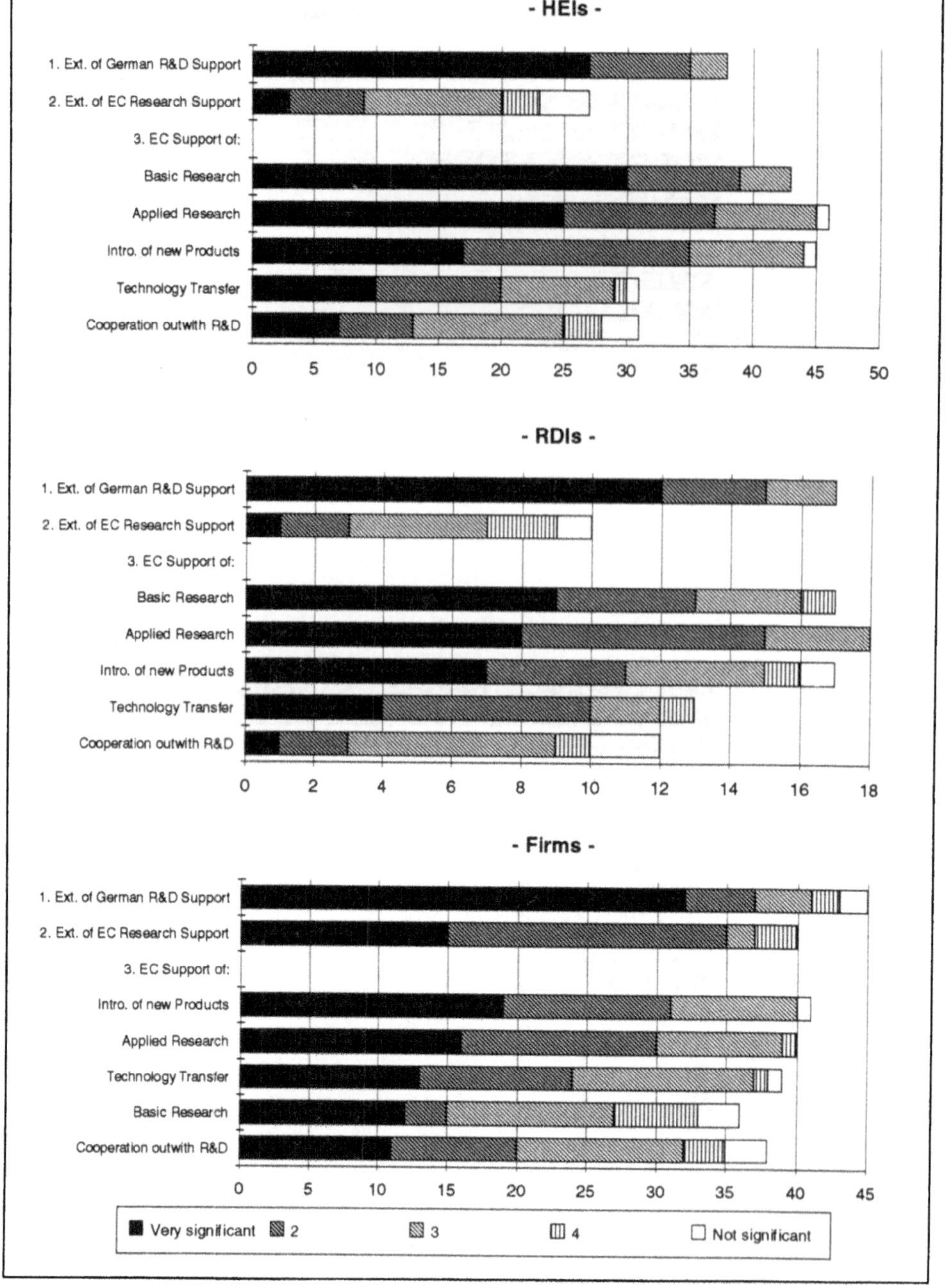

and RDIs. The support of market introduction and technology transfer is mentioned often by enterprises in the new Länder; this shows the firms' problems in introducinng products onto the market. These answers reflect the situation in the new Länder, where at present R&D financing is very heavily dependent on federal support and the companies' activities are short-term oriented. National support is regarded as the primary source of support, but great hopes are placed in EC support for future perspectives.

7.6 Summary and Conclusions

There are potential future participants in EC RTD support programmes in the new Länder; this emerges clearly from our questionnaire. Only two institutions from the new Länder took part in the Second Framework Programme, due to the fact that the integration of the former GDR into the European Community took place at a time (1990) when this Framework Programme was already far advanced. However, there are other important **barriers and limitations to participation** which in the future will have to be progressively dismantled:

- The **restructuring of the R&D infrastructure** (in industry and the public sector) since 1989 has been so drastic that the actors are now in a phase of slow consolidation in which they are gradually recuperating their full powers of action.

- Many sectors of industry in the new Länder are in a severe state of crisis, giving rise to feverish short-term activity and making it difficult to follow long-term research and innovation strategies. For a long time to come, most East German enterprises will be interested mainly in **market-oriented support**.

- Actors in the new Länder have **lost most of their cooperative links with Eastern Europe** (structural changes also in Eastern Europe; collapse of market relations) and **still have to build up relationships with partners in Western Europe**. In this respect, their situation is comparable with that of institutions in Western Europe before the support of transnational R&D cooperations by the European Community was introduced. Research capacities in the new Länder need help in their process of integration into the European Community, and help in reactivating useful links with Eastern Europe.

The situation in which the new Länder are forced to make a new beginning potentially holds **great opportunities**: since accession to the Federal Republic of Germany and integration into the European Community have taken place at the same time, it should be possible - as well as the necessary "anchoring" within the national system - for European and national research cooperations to be initiated and developed in parallel right from the start.

8. Assessment and Conclusions: Impact of Community Policy for the German Research Landscape

In the preceding chapters we have presented the results of our written questionnaire and interviews with R&D performers in the German research landscape. This chapter gives a summary and assessment of these results as we see them. The structure of the chapter does not follow the various groups of protagonists separately, but centres on a cross-group examination of the following points:

- the **importance** of EC RTD support for the German research landscape and for participating R&D institutions,
- a **characterisation** of EC projects,
- the **aims** of the Second Framework Programme and the extent to which they are reflected in the aims of participants,
- **impacts** on R&D performers and policy actors, and their relationship to the aims of the Second Framework Programme.

8.1 Importance and Scope of Community Technology Policy

8.1.1 Importance for the German Research Landscape

Awareness of the possibilities of EC RTD support varies between the different groups of research actors in Germany. In industry, it is above all large companies in the sectors of electrotechnics, information and communication technologies and aerospace that are highly informed. Small and medium-sized firms, and large enterprises (BIGs) in the chemical industry and mechanical engineering, which play an important role for the Federal Republic of Germany, participate less in EC projects and are insufficiently informed on the possibilities of EC RTD support. Among the non-university R&D institutes (RDIs) the Fraunhofer Society, the Max Planck Society and the national research centres are particularly well-informed on EC support. At the higher education institutes (HEIs), awareness of EC support possibilities is strongly dependent on the activities of the particular institute or research group concerned. All in all, **science actors have be-**

come more aware of EC support over the last few years: their institutions are appointing EC experts, setting up offices for EC representatives and have initiated the KoWi, the Liaison Office of the German Scientific Research Organisations (Koordinierungsstelle EG der Wissenschaftsorganisationen), with offices in Brussels and in Bonn.

A background factor influencing the situation at present is that federal and Länder funding for research is stagnating. For the next few years at least, the public sector will have to work under restrictive financial policy conditions (BMFT 1993, 7). Actors in the German research landscape are trying to compensate this development by stronger participation in EC projects, and they expect EC funding to increase in the future. However, this hope of **compensation** is deceptive: on the one hand, the balance between German contributions to the EC and the reflux of funding from the European Community to German participants ("juste retour") is now almost in equilibrium; a substantial increase is not to be expected. On the other hand, EC funds for research are also limited; unless they are increased, a growing number of applications will only lead to higher refusal rates in the various EC programmes and to increased complaints about rejected proposals. Expectations of an over-proportional increase in RTD funding of the CEC are uncertain in the present economic and political situation of Europe.

The **financial share of EC RTD support** in total expenditure on R&D in Germany is small. In the period from 1987 to 1991, EC research funding amounting to ca. 1.3 thousand million DM was received by researchers in the Federal Republic of Germany. This amount corresponds to 1.8% of total R&D expenditure by the German Federal Government for the same period. Compared with the total expenditure by all German research actors (domestic only) for the same period, the EC contribution represents only ca. 0.4%. In assessing the significance and impact of EC research support for the German research landscape, this limited financial proportion should always be borne in mind, in order not to overestimate the possibilities of EC support.

Viewing the German research landscape as a whole, German participants in EC programmes form a relatively **"small community"**. There were 2,108 German participations in EC projects under the Second Framework Programme. The number of innovating SMEs in the Federal Republic of Germany is estimated at 25,000 (BMFT 1993, 105); there were only 166 EC project participations by SMEs. The number of EC participants is substantially reduced if the frequency of participation is taken into account: the "top ten" large enterprises (most frequent participants) accounted for more than half of all participations in the group of large companies, and the "top ten" HEIs for somewhat less than half of all participations in the higher education sector. German participants in the Second Framework Programme obviously form a stable clientele who - once they have entered the circle of EC consortia - cooperate several times on different projects. To newcomers (especially SMEs and higher education institutes), access to EC projects appears difficult. EC projects are occasionally described by non-participants as a **"closed shop"**.

The German research landscape still has a **large potential for future participation** in EC RTD support. This applies particularly to small and medium-sized firms, institutes and research groups at higher education institutes, private R&D institutes, and to some extent to large companies in industrial sectors which have not been directly addressed by EC programmes so far. **Reasons for non-participation** are to be found primarily in information deficits about the possibilities of EC support, in the expectation of too much administration in the project, in less prevalent European orientation, but also evidently in satisfaction with German RTD support up till now.

The fact that **participation by SMEs** in the Second Framework Programme is still **small** can be attributed on the one hand to problems in gaining access to "networks", the limited R&D resources, and the complicated and time consuming application for EC projects. On the other hand the specific innovation pattern of these firms is oriented to marketable products and less to applied research. The precompetitive RTD support in the Second Framework Programme is oriented primarily towards the innovation patterns of large companies and the research aims of HEIs and RDIs. If policy is aiming at a stronger participation of SMEs, it will have to take account of the existing division of labour between large enterprises and SMEs in the macroeconomic innovation process. Since, however, Community RTD policy wishes not only to accelerate the generation of new technologies, but also to speed up their diffusion throughout Europe, and since SMEs are increasingly involved in this diffusion process through international activities, they represent an important target group. For a stronger participation of SMEs, special policy instruments are needed which are oriented towards easier application procedures, transnational cooperation, development activities and the diffusion of technologies.

The new Länder were not affected by the Second Framework Programme. Nevertheless, it emerged clearly from our enquiries that there are great expectations of Community support there. R&D institutions in the new Länder have special problems which consist especially in language deficits and a lack of management experience, and EC advisory offices in the new Länder should take these aspects into account. For the time being, however, the special orientation of innovative behaviour (re-structuring within firms, marketable products, predominance of product development and diffusion, short-term orientation), and the orientation towards federal German funding, mean that the EC participation possibilities of R&D performers in the new Länder are more limited; this will certainly change within the coming years.

8.1.2 Strategic Importance for Participants

The EC projects are an **integral part of the R&D portfolios** of participants. With the exception of some of the large companies, projects belong primarily to the **strategic core area of technological competence** (cf. Figure 8.1.2.1). It is mainly the large companies and RDIs that are aiming to open up a potential new area. For all participants, the EC projects serve to **extend existing business or research fields**, and not so much to establish new ones. Only large companies use projects to reduce costs; diversification of business fields is considered less important. The EC projects have an integral place among the R&D projects of the participants, and play an important part in extending their technological competence.

Higher education institutes and R&D institutes oriented towards basic research (Max Planck institutes, national research centres, "Blue List" institutions) mainly apply to participate in EC projects when the topics fit into their given research context. Large enterprises, and to some extent SMEs, would in some cases have carried out their R&D project without EC support, but in a much less extended form and over a longer time period. The incorporation of the EC project into the strategy and thematic fields of R&D institutions certainly means that the project is useful for them, but reduces the **"value added"**. An absolute "value added" is not attainable in practice; the EC projects arise from existing R&D projects or programmes and deal with related problems.

The **technological level of EC projects** is assessed ambivalently by large companies: a proportion of them (not to be underestimated) only attaches a limited importance to EC projects in the extension of technological competence, and describes the technological level as average (cf. Figure 8.1.2.1). Interviews with BIGs confirm this divergent assessment of EC projects.

The financial significance of EC support is described by BIGs as small; it is - as one R&D head put it - "a drop in the ocean". For higher education institutes, R&D institutes and SMEs, however, EC funding may amount to between 10 and 20% of their R&D budget. For the SMEs, the average share of EC funding is actually higher than the share of federal funding; this reflects the fact that EC projects are often too large, especially for small firms. Generally speaking, EC support mainly enables the participants to carry out R&D projects **in an extended form** and **within a shorter time**.

With the exception of SMEs, all other groups in German research expect the **financial importance of EC support to increase**, and Federal and Länder support to decrease (cf. Figure 8.1.2.1). Thus, great expectations are being placed in the Commission of the European Communities.

Figure 8.1.2.1: Strategic Significance of EC Project

	SMEs	BIGs	HEIs	RDIs
R&D Project Portfolio				
Integrated part	very important	very important	very important	very important
Few connections	important	important	less important	important
No connections	not important	not important	not important	not important
Role of Technology Competence				
Strategic technological area	very important	important	very important	very important
Potential new area	less important	important	less important	important
Just one technology among others	important	very important	important	important
Rather peripheral area	not important	not important	not important	not important
Role for Business Strategy				
Establish new research areas	important	important	important	important
Extension of a research area	very important	very important	very important	very important
No significance	not important	not important	not important	less important
Diversification	less important	less important	not important	less important
Cost reduction	not important	important	not important	not important
Role of EC Funding				
Independent of EC funding	less important	less important	not important	not important
Only with EC funding	very important	important	very important	very important
Larger scale possible through funding	important	very important	important	important
Faster through funding	important	important	less important	less important
Other goals without funding	less important	less important	not important	less important
R&D Budget				
Share of EC funding	less important	not important	less important	less important
Share of Federal Republic funding	not important	less important	important	important
Future Financial Significance of ...				
EC R&D funding	→	↗	↗↗	↗↗
Contracts from enterprises	↗	↗	↗	↗
R&D funding of Federal Republic/Länder	→	↙→	↙→	↙→

●	very important	→	constant significance
◍	important	↙→	constant, rather decreasing
⊜	less important	↗	increasing significance
○	not important	↗↗	strongly increasing significance

8.2 Character and Aims of EC Projects

8.2.1 Character of EC Projects

Regarding their research typification, EC projects can be characterised as **applied research**; enterprises also use them for experimental development, and for SMEs demonstrators and diffusion are important aspects (cf. Figure 8.2.1.1). The research-intensive HEIs and RDIs carry out basic research in the EC projects; for them, experimental development is either unimportant or not very important.

On the one hand, this result indicates that German participants have anticipated the aim of the Second Framework Programme to promote precompetitive research activities; on the other hand, it expresses the division of labour between participants: industrial participants are more strongly oriented towards applied R&D activities in the projects, whereas universities and institutionally financed R&D institutes perform more basic research and provide expert knowledge within the consortium.

In industry EC projects are directed towards a **medium-term exploitation of results within about three years** after completion of the project. For SMEs, projects with a shorter time horizon are also important, and for BIGs, projects with a longer time horizon (cf. Figure 8.2.1.1). For the **HEIs and RDIs**, EC projects have a **long-term** orientation, for **contract research institutes** a medium-term orientation.

Large enterprises, higher education institutes and R&D institutes tend to cooperate mainly in **large consortia** with more than five participants. **SMEs** tend to cooperate in **small EC projects** with an average of up to five partners (cf. Figure 8.2.1.1). The **cooperation behaviour of participants in the consortium** is problematic: enterprises aim primarily at cooperation with other firms in the EC, the higher education institutes want to cooperate with (European) HEIs. Orientation towards, and cooperation with, partners in other groups is considered less important. To some extent at least, RDIs constitute an exception: they seek cooperation both with universities (this applies particularly to institutes performing basic research) and with enterprises (sought especially by contract research institutes).

Figure 8.2.1.1: Character of EC Projects

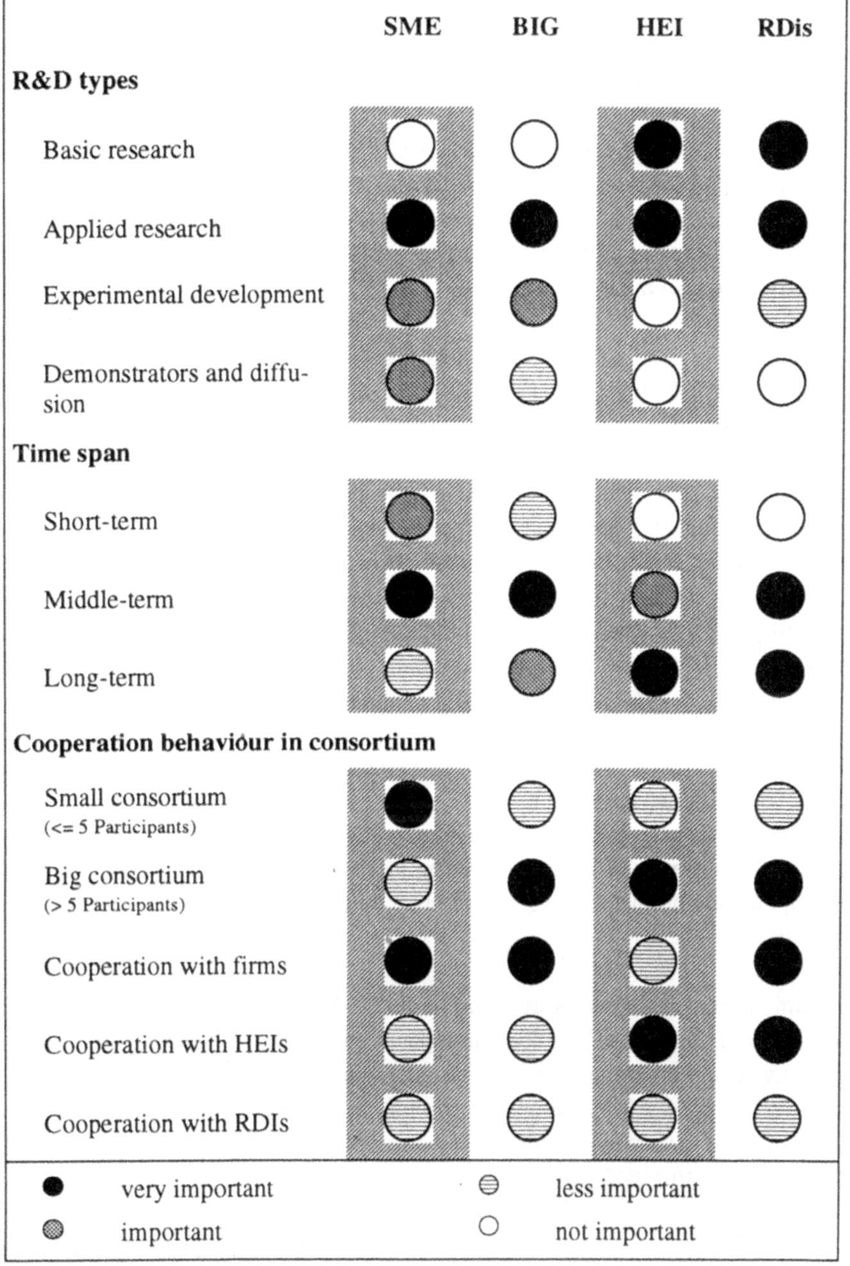

8.2.2 Correspondence of Support Aims and Participants' Aims

With the Second Framework Programme, the European Communities are pursuing the following primary aims:

I) to strengthen the **scientific and technical basis** of European industry,
II) to support the development of **international competitiveness,**
III) to support **transnational cooperation** between industry and science, and
IV) to give impetus to the concept of a Single Market by the development of **standards and norms.**

The intention was to achieve these aims mainly by promoting **R&D activities in the precompetitive field**. To what extent are these policy aims reflected in the motives for participation in EC projects, and in the aims of the individual projects ?

In the EC projects German participants (the enterprises, too) conduct mid-term and applied research and contribute essentially to the main aim I of the EC support, to **strengthen the scientific and technical basis** of European industry. In their EC projects the German participants are all pursuing the aim of extending their knowledge base and improving their qualifications (cf. Figure 8.2.2.1 and 8.2.2.2). The higher education institutes and R&D institutes are also aiming to enhance their scientific reputation; the HEIs are using the projects for the preparation of PhD theses.

Economic aims play a **smaller role** for participating German enterprises than do the enhancing of skills, an orientation towards product innovations is only mentioned in every second case. Much more importance attaches to the aim of new or improved products than to new or improved process technologies (cf. Figure 8.2.2.2). The aim of developing new products is even more important for participating SMEs than for large enterprises.

The aim of **transnational cooperation** in the EC between industry and science is anticipated by nearly all of the German participants. In the **project aims** a **lack of cooperation** between actors in industry and science is sometimes apparent already in the orientation (cf. Figure 8.2.2.2). European cooperation in the field of R&D is thus a very important motive for participation in EC projects; however, there is only a limited relationship between partners of different sectors.

According to statements by industrial participants, **explicit technical goals** such as the building of prototypes or demonstrators, the development of methods and instruments

Figure 8.2.2.1: Why do you carry out R&D Projects with EC Support?

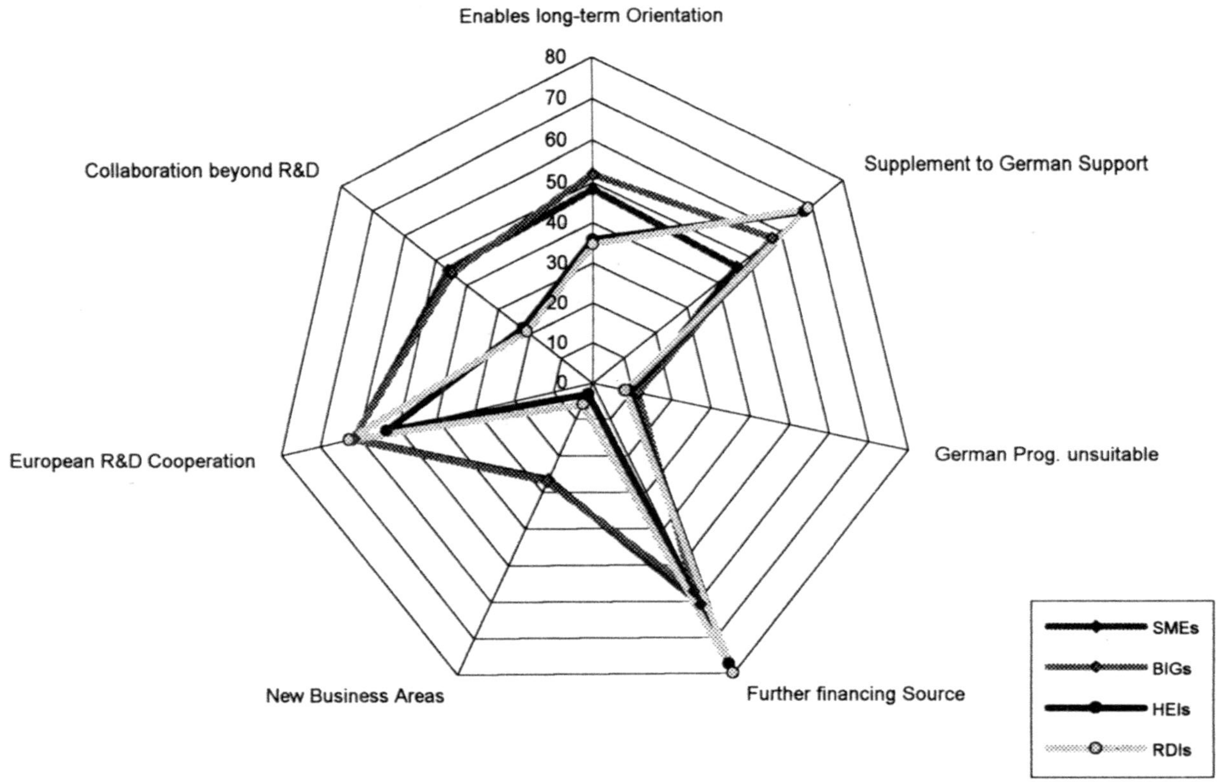

Figure 8.2.2.2: Which of the following Goals did you aim for in this Project?

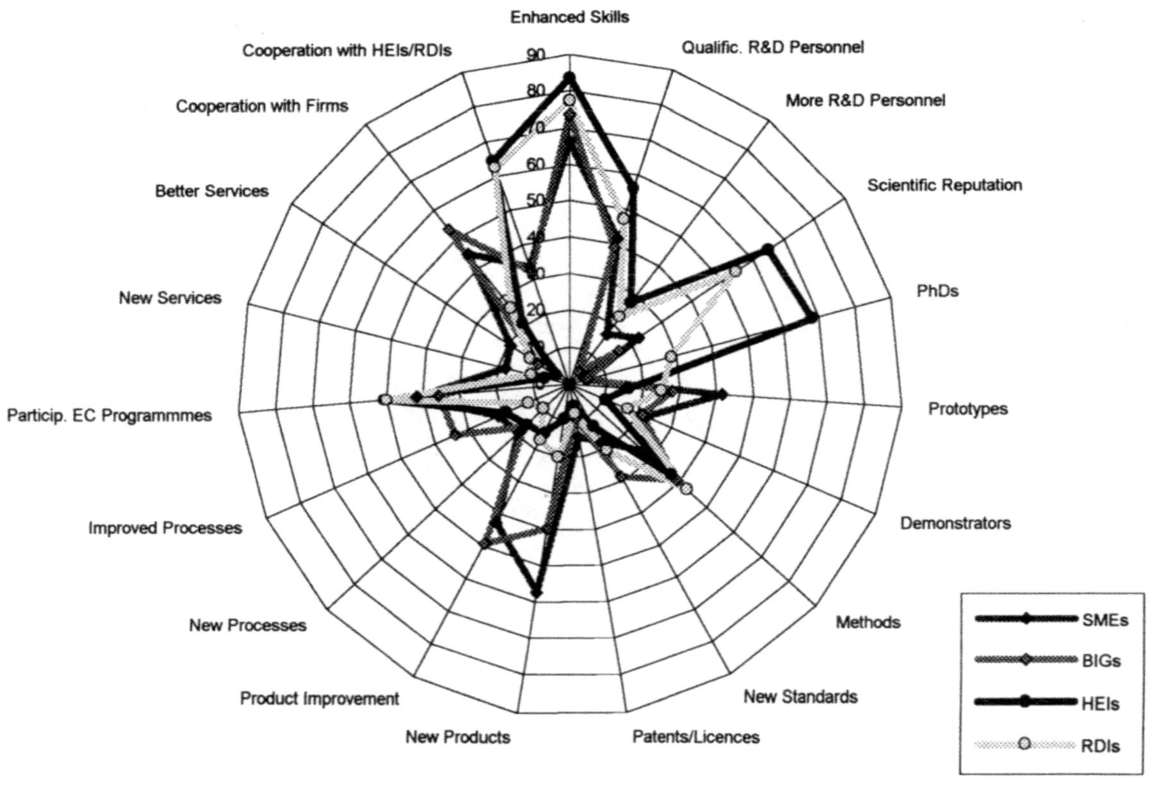

are rarely being pursued (cf. Figure 8.2.2.2). In this context it is noticeable that the **development of new standards,** considered by the Commission of the EC as important for the concept of a Single Market, is envisaged by German participants **to a limited extent only**.

8.3 Results and Impacts on German Actors

8.3.1 Improvement of the Scientific and Technical Basis

The most important results (cf. Figure 8.3.1.1) and impacts (cf. Figure 8.3.1.2) of EC projects are **improved knowledge and scientific skills**; this is equally true for all German participants. The higher education institutes also make use of EC projects to increase their R&D personnel, to enhance their scientific reputation and for PhD research (cf. Figure 8.3.1.1). Viewed as a whole, the success of the Second Framework Programme lies in having promoted the exchange of ideas, technology and knowledge within the European Community and in having improved the scientific and technical basis.

In the course of the Second Framework Programme the **scientific and technical quality of the results**, sometimes considered unsatisfactory in the 1980s, has been acknowledged by all groups of actors in German R&D. In some areas of technology, according to experts, leading positions have been achieved; however, in the field of information and communication technologies these successes are limited and cannot be interpreted as signs of a broader breakthrough in the improvement of international competitiveness.

EC projects provide opportunities for HEIs and RDIs to extend their research activities in fields where they are already active. The relatively low participation of HEIs in the Second Framework Programme can be attributed, among other causes, to the good quality, the relatively good financial equipping and the functional autonomy (through the German Research Association, DFG) of the German support system. Institutional funding, and the distribution of funding via largely autonomous science organisations, provide a guarantee for high quality in science and research in Germany. R&D actors fear that in future, if federal funding is in short supply, they will be **confronted with the political argument** that deficits in resources should be compensated by the acquisition of EC funding. However, due to the small relative volume of EC funds, this will not be possible.

Figure 8.3.1.1: Which Results did you achieve in this Project?

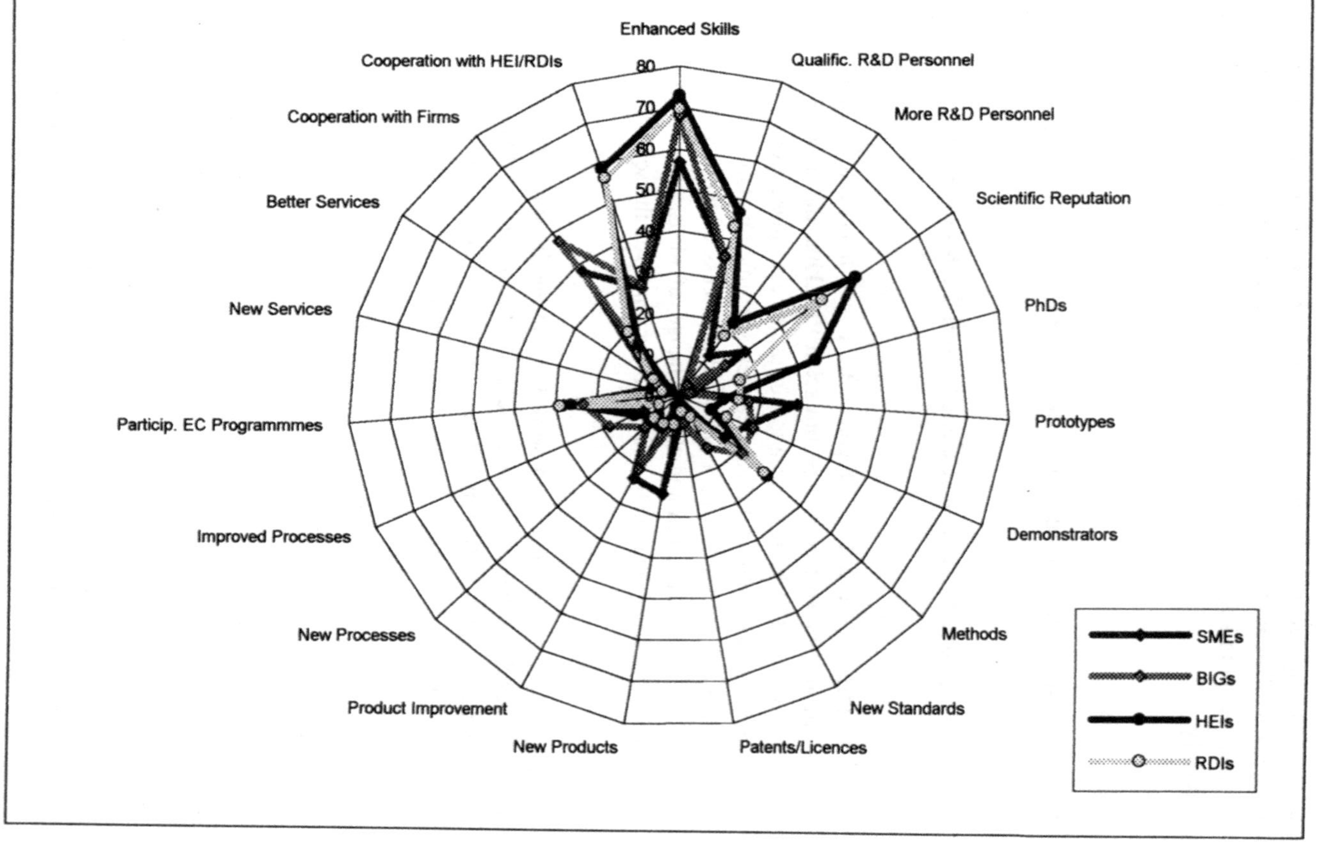

8.3.2 Impacts on Transnational Cooperations

One important effect of EC support in the sense of politically promoted "value added" is the increased willingness of all participants to cooperate with other institutions throughout Europe (cf. Figure 8.3.1.2). The **important aim of bringing European actors together** and creating awareness of a **"European technology community"** has been attained by the Second Framework Programme, if one considers the statements of German participants. The EC programmes have succeeded in dismantling a previous reluctance on the part of industry to work with partners from other countries in strategic core technologies. For firms in Europe, the idea of cooperating with one another as partners has definitely become more attractive.

Two points relating to cooperation appear to be problematic, however: firstly, the **desired cooperation between industry and science** takes place **only** to a **limited** extent. Secondly, it is noticeable that the wish to engage in further cooperations **definitely focuses on cooperation within other EC projects**; the intention to cooperate in transnational R&D projects that are not supported by the EC is expressed by only 10% of German participants (cf. Figure 8.3.1.2). One fifth of SMEs, and one sixth of large enterprises, have engaged in cooperation on the development of joint products; only in a few isolated cases up till now have EC partners cooperated in production, marketing or sales. Obviously - and this also emerged from interviews with enterprises - R&D goals are pursued jointly, but the results of EC projects are used and exploited separately.

8.3.3 Impacts on Strategies and Competitiveness

As already suggested (particularly in Chapter 4), it is too soon to assess the positive impacts of the Second Framework Programme on the international competitiveness of German industry (cf. also Krull 1993, 29ff). Enterprises indicate that they expect their project results to be economically exploitable within three years; an additional period of time is needed before a product innovation results in a growing market share or an improved balance of trade for this product group. A consideration of the issue whether, from the viewpoint of German industry, the Second Framework Programme has led to an increase in national competitiveness has to be **limited to two aspects**: firstly, the question whether the **structural basis** for competition has been improved, and secondly whether the **strategies of enterprises** have been changed, with noticeable effects.

Figure 8.3.1.2: Which Impact has this Project had on your Organisation / on your Cooperation with the European Partners?

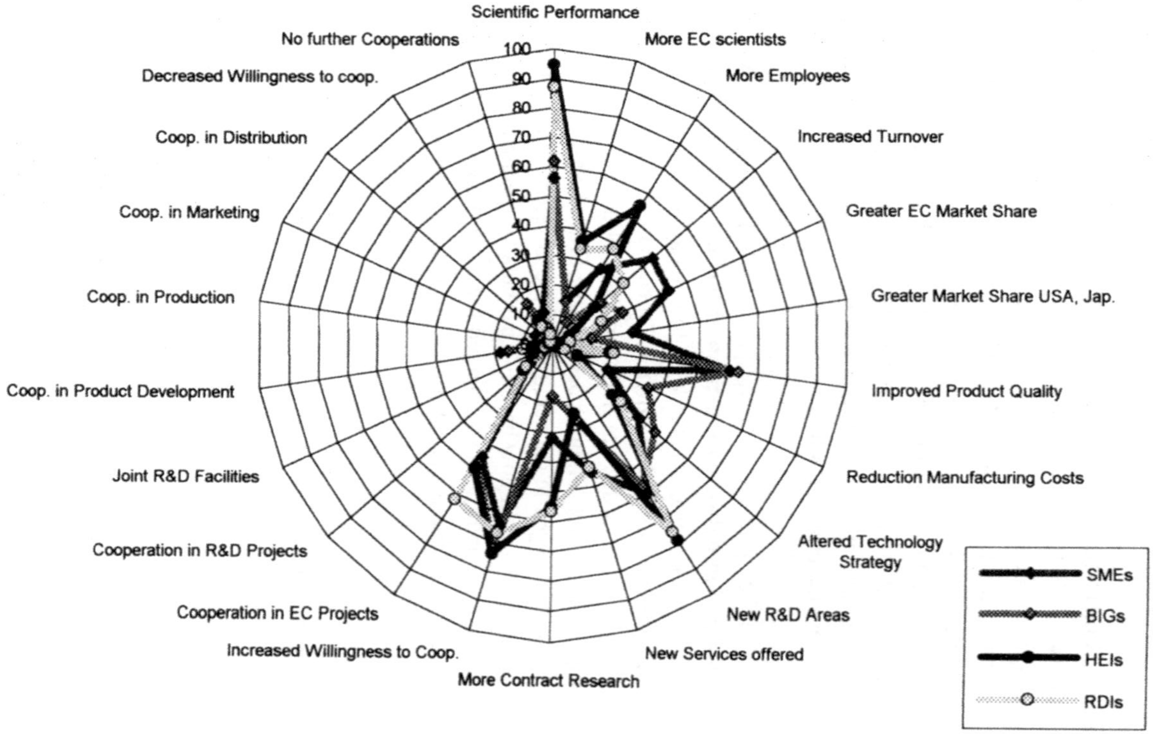

The economic goals of German firms tended to relate to new or improved products; only in a few cases were process innovations being aimed at (cf. Figure 8.2.2.2). This is apparent in the economic impacts, especially in the achievement of higher product quality; more than 60% of the enterprises have attained improvements here with the EC project (cf. Figure 8.3.1.2).

Preliminary stages of product or process innovations such as prototypes, demonstrators and the development of methods/instruments were only pursued - and attained - in a relatively small number of cases (cf. Figure 8.3.1.1). Compared with the British IMPACT study (PREST/SPRU 1992) there is a markedly **different pattern of responses from participating firms in Germany**: although aims such as enhancing skills and knowledge, and further participation in EC projects, rank high among the aims achieved by British firms, they are closely followed by economic results such as new process technologies, new products, the developing of prototypes or the improvement of existing products.[36] This result, taken in conjunction with the characterisation of EC projects by participating German firms, means that participants in German industry are using EC projects more than British firms for their medium-term research activities and for the extension of their science and technology base.

Changes in the strategies of firms towards a greater willingness to cooperate have already been discussed. However, EC projects have,also **opened up new research fields**, and in some cases have changed the technology strategy of firms (cf. Figure 8.3.1.2). Remarkably, the **achievement of patents and licenses was hardly ever aimed at**. This should be borne in mind, because patent applications are often taken by the Commission as an indicator for the success of EC projects.

If enhanced scientific skills, the acquisition of know-how, transnational cooperation and changes in technology strategy are regarded as factors which improve competitiveness, it can be stated that the Second Framework Programme has **improved the basis of international competitiveness for participants in German industry.** Nevertheless, this companies mainly the German and European markets and less the Japanese or US mar-

---

36 In answer to the question, which results were achieved or expected from the EC projects, the **British enterprises questioned** replied (not a complete account): enhanced scientific skills (75% of the answers), further EC projects (65%), new process technologies (55%), new products (51%), prototypes (49%), improved products (48%), demonstrators (45%), improved process technologies (42%), standards (30%). The **German enterprises** replied to the question, which results were achieved within the EC project or were expected within 3 years after project end in the following manner (not complete account): enhanced scientific skills (67% of the answers), further EC projects (31%), new process technologies (13%), new products (37%), prototypes (28%), improved products (38%), demonstrators (20%), improved process technologies (21%), standards (18%).

kets. However, the efforts, and successes, in the field of norms and standards are smaller than expected.

To complement the analysis of the scientific, technical and economic impacts of EC support, German participants were also asked about generally positive **social and environmentally relevant impacts** of EC support. However, these general impacts **play only a limited role** in the answers. Positive impacts on environment, health, cultural integration and energy conservation were mentioned by the participants; however, these are attributed mainly to projects whose aims explicitly related to these topics.

8.4 Impacts on Policy

One of the most important ways in which the **European Community influences German RTD policy** lies outside the EC Framework Programme being considered in this study: the **"Community Framework for State Aids for Research and Development"** (cf. Commision of the European Communities, 1986) stipulates from a competition policy perspective a notification of all R&D support measures of Member Countries according to common criteria. Therefore, science-oriented basic research may be financed 100% from public sources, industrial basic research up to 50%, and applied research from 25% up to a maximum of 50%. This has already **noticeably reduced the freedom of action of the German actors** in the field of industrially-oriented RTD policy, namely the BMFT, the BMWi and the "Länder" governments.

EC RTD policy in the narrower sense, as expressed in the **Framework Programmes, gained influence** in Germany slowly at first, then with increasing momentum: up to the mid 1980s, it was regarded as a marginal factor; with the Second and particularly the Third Framework Programme it has become a **recognised** actor in the German research landscape, **although not yet** of **central** importance. The impacts of Community support have mainly been in the stimulation of original, **additional research and development efforts** in industry and in the industrially-oriented research system. This has taken place over the past few years in the context of an evolving (informal) **division of tasks** which - although tentative and lacking in stability - has been mainly **accepted** by the actors involved:

- the **EC** concentrates its programmes on **transnational research tasks**, the creation of a Single European Market and on improving the competitiveness of European industry; this includes the transnational concentration of expert knowledge, cost saving through cooperation on high-cost projects and the joint elaboration of EC norms and standards.

- The **Federal Ministry for Research and Technology** (BMFT) concentrates on supporting **basic research, preventive research**, and **technology developments** with a long time horizon. The **BMFT,** the **BMWi** (Federal Ministry for Economic Affairs) and the **Länder governments** carry out measures for the support of **innovative capabilities in industry** from a national and regional viewpoint respectively. The **Länder** governments are responsible for safeguarding and promoting science in the **higher education sector** in coordination with, and sharing some tasks with, the Federal Government.

What falls within the **transnational sphere of the EC**, and what does not, is to some extent **open to interpretation** and **subject to political negotiation processes**. This is particularly apparent in the strong commitment of EC RTD policy to information technologies; it is not possible to decide here whether this commitment is the outcome of the perception of a transnational task, since it is only by joining forces that European industry can keep pace with worldwide developments in this field, or whether it is simply the consequence of a financially motivated failure by the Member States to perceive a national task.

The **considerable value added** of Community support in Germany concerning impulses for additional or new R&D activities that emerges from our analysis of the Second Framework Programme, and that can also be assumed for the beginning of the Third Framework Programme, has been **called into question** since the unification of Germany, and noticeably **since 1991**: as a consequence of significant bottlenecks in the domestic budget, German public RTD support has been stagnating; there have been substantial decreases in support funding in central areas, such as the research support of the DFG. It is probable that German researchers have begun to seek **other sources of financing**. Since this time, the possibilities of **EC research financing** have appeared in a new light; however, total compensation via the Community is neither desirable nor realistic.

Compared with the governments of other Member Countries, participation of the German Federal Government in the shaping of EC RTD policy has been very restrained. Only in the last few years has it asserted its influence, and begun to take seriously the coordination of Community measures with national RTD policy. The **Länder governments** play a special role for the German RTD policy; according to Article 30 of the "Grundgesetz" - the German Constitution - the Länder are responsible for science and research. Over the previous decades, a tentative **distribution of tasks** has arisen in the RTD policy efforts between the Federal Government, increasingly strong in research policy, and the "Länder", whose resources since the 1970s have been taxed to the full by the rapid expansion of the higher education sector. This working arrangement now appears to be endangered by the coordination requirements of EC RTD policy: the Länder see themselves being permanently affected by decisions in Brussels, but **do not feel** that they are **adequately represented** there. Although they do now have their own representative (representation is limited to observer status), he (or she) is faced with the problem of representing the interests and positions of 16 different "Länder". In view of this structure of the representation of "Länder" interests in Brussels, the occasionally voiced "sandwich theory" (according to which the Federal Government is losing influence, caught between a powerful EC policy on the one side and Länder governments gaining strength on the other) lacks factual foundation.

The **large science organisations** in Germany (in particular the German Research Association, the Max Planck Society, the Conference of University Rectors (Hochschulrektorenkonferenz) and the Science Council) fear that if EC RTD policy gains further importance, the **tried and tested principles and mechanisms of self-organisation in German science**, which are **not found** in the same form in other European countries, will be undermined, particularly since typical EC project support programmes, similarly to the project support of the BMFT, are shaped far more by "political" criteria than institutional funding, and are not predominantly determined by inherent scientific criteria. In order to counteract this process, they suggest that the principles of self-organisation in the science system should be established if possible **throughout Europe**. The most important suggestion made in this respect is to set up a **"Science Foundation on Community level"**, on the one hand to approve projects in basic research with the assistance of expert and programme committees and to allocate funding, and on the other hand to advise the EC, in joint committees with representatives from industry, on the conception of industrially-oriented programmes for the support of application-oriented research.

In view of the low approval ratio of many EC programmes, RTD policy actors in Germany complain that the **programmes** are **not sufficiently focused** and the calls for projects not sufficiently specific. This gives rise to a "flood" of applications, which in principle could be avoided.

Many RTD policy actors in Germany fear that as Community support gains importance, there will be too high a **concentration** of related tasks and competences within the **policy administration system of the EC**. They are afraid that the planning, decision-making and control processes of EC research administration, although acceptable in themselves, may become untransparent and eventually ineffective. Against this background there are calls for **decentralisation** of the institutions and processes of European research and technological development policy. However, it is still an open question whether this decentralisation regards only the administration of (sub-)programmes, or whether it also includes the processes of developing an informed opinion prior to policy decisions.

In Germany, where the various actors of the research landscape initially made only hesitant use of the possibilities of European RTD support, there has arisen over the last five years a **broad and differentiated infrastructure** of institutions which have the task of **bringing researchers into contact with EC support**: as well as the industrial lobbyists in Brussels, the KoWi, the EuroInfo Centres for industry, the EC advisory offices of the chambers of industry and commerce, the EC representatives at HEIs and the contact offices of the Länder governments in Brussels, these include a number of other attempts to improve the coordination between Community administration and its clientele. One of these is the setting-up of EC contact offices at research institutes with experience of EC support schemes, with the aim of improving the flow of information and feedback between (potential) clients and EC administration (as in Baden-Württemberg). Another attempt is the setting-up of the Value Relay Centres, which work together in a network all over the Member States.

It is conceivable that **both tendencies** - the almost inevitable **decentralisation** of part of the Community apparatus on the one hand, and on the other the growing infrastructure of **institutions for the support of interaction** between the research landscape and the EC, may eventually **grow together** to form a basic element of a new European institutional structure of research and technological development support.

9. Outlook

Important reasons for the present high degree of acceptance of EC support in Germany are the **transnational cooperation** towards which the EC programmes are directed and the orientation towards **medium-term or applied research.** Provided the participating enterprises do not have to worry about distortions of their competitive situation or know-how "leaks" to competitors they will participate in a joint R&D project. Thus, from the viewpoint of many German actors, **the orientation of the Community programmes** towards applied research is a "conditio sine qua non" for the shaping of future EC programmes.

In the future, the EC policy guideline of **"strengthening the international competitiveness of European industry"** will have to be understood in a broad and **problem-oriented** sense: technology as the basis for industrial innovation and competitive production at the beginning of the 21st century cannot be understood or dealt with from a conventional standpoint. **Various fields of technology are progressively fusing and overlapping**. For instance, in order for nanotechnology to function as a fruitful base technology for future innovation processes and new generations of technology, interdisciplinary interactions are necessary with electronics, information technology, materials science, optics, biochemistry, biotechnology, medicine and micro-mechanics. The applications of nano technology extend into the fields of "tailored" materials and biotechnical systems, but are mainly in electronics. The science-based technology of tomorrow requires the sustained support of appropriate fundamental research. Not only does this increase the importance of the classic ***transfer from basic research*** to industrial research; the feedback transfer from industrial problems ***into basic research*** also acquires new significance.

Even more important is the resulting challenge to the competitiveness of industry. The **path of development followed by the developed industrialised nations** until now **cannot be continued** in the long term, due to the consumption of **resources**, the damage to the **climate** and the dwindling of **employment**. Public technology policy will have the important task of preparing the "turning of the ways" into acceptable paths. This implies the necessity to link **visionary applications of new technologies** with **new research tasks**, without eroding the present responsibilities of the actors. Public technology policy can make an important contribution by an intelligent combination of "classic" research support, stimulation of demand, the establishing of appropriate frame conditions and stable, long-term signals to science and industry. The **global nature of ecological and social challenges** will mean that the Commission of the European Communities, as a transnational authority, will be pressed by a growing number of urgent problems that can **no longer be dealt with at the level of** small and medium-sized

countries as in Europe. The transnational share in the shaping and planning of research and technology will continue to grow. The sooner the Commission, in cooperation with the Member States, recognise and acknowledge this challenge, the sooner it can be met, thus making a contribution towards **strengthening the competitive situation** of Europe on the world markets.

To achieve this, Community policy in future will have to promote not only R&D cooperations within Europe, as it has done until now, but will have to specifically envisage the support of **cooperation with partners in the "triad"** (USA, Japan) and with **developing regions**.

The present study has shown that - particularly in the last few years - expectations in Germany regarding EC research and technological development policy have markedly increased. German actors assume that by now EC structures in research and technology have been functioning for a long time according to **their own dynamics**. This **development is no longer reversible** - the issue now is a responsible shaping of developments.

References

BMFT (Bundesministerium für Forschung und Technologie) (1990):
Faktenbericht 1990 zum Bundesbericht Forschung 1988, Bonn

BMFT (Bundesministerium für Forschung und Technologie) (1993):
Bundesbericht Forschung 1993, Bonn

Bräunling G./ Maas, M. (1988):
Nutzung der Ergebnisse aus öffentlicher Forschung und Entwicklung, Forschungsbericht an die EG, Karlsruhe

Commission of the European Communities (1986):
Official Journal of the European Communities, 86/C 83/02 of 11.4.1986, Brussels/Luxembourg

Commission of the European Communities (1990c):
DRIVE: Progress Report and mid-term Review. SEC (89) Final Report 1990, Brussels

Commission of the European Communities (1993):
Evaluation of the BRITE/EURAM Programme (Areas 1 to 4) - Final Report, Brussels, February 1993

CREST - Ausschuß für wissenschaftliche und technische Forschung (1992):
Bewertung des zweiten FTE-Rahmenprogramms. Bericht des CREST an den Rat, Brüssel

Ergas, H. (1987):
Does Technology Policy Matter? In: Guile, B.R./Brooks, H.: Technology and Global Industry. Companies and Nations in the World Economy, Washington D.C., 191-245

Eurostat (1992):
Regions - Nomenclature of Territorial Units for Statistics (NUTS), Brussels

ISI/IWH (1993) (Kuhlmann, S./Berteit, H./Bierhals, R./Holland, D./Reger, G./Adam, E./Röhnert, S.):
Wirtschaftsnahe Forschungseinrichtungen in der Bundesrepublik Deutschland unter besonderer Berücksichtigung struktureller Anpassungen in den neuen Bundesländern. Studie im Auftrag des Bundesministers für Wirtschaft. Berlin/Karlsruhe

Jäger, F. (1992):
Planning and Reviewing the Participation in European R&D Programmes: the Experience of Siemens. In: Research Evaluation, Volume 2 Number 1, April 1992, Guildford (Surrey)

Kommission der EG (1987):
Amtsblatt der Europäischen Gemeinschaften L302 vom 24.10.1987, 30. Jahrgang, Brüssel/Luxemburg

Kommission der EG (1990a):
Amtsblatt der Europäischen Gemeinschaften L117 vom 08.05.1990, 33. Jahrgang, Brüssel/Luxemburg

Kommission der EG (1990b):
Forschungs- und Technologieförderung der EG, 2. Auflage, Brüssel

KoWi (Koordinierungsstelle EG der Wissenschaftsorganisationen) (1992):
Forschungs- und Technologieförderung der Europäischen Gemeinschaft: Ziele Instrumente, Prinzipien, Brüssel/Bonn

Krull, W. (1992):
Neue Strukturen für Wissenschaft und Forschung. Ein Überblick über die Tätigkeit des Wissenschaftsrates in den neuen Ländern. In: Aus Politik und Zeitgeschichte; Beilage zur Wochenzeitung das Parlament, B51/92, 11. Dez. 1992, 15-28

Krull, W. (1993):
Guidelines for the Evaluation of European Community R&D Programmes, Brussels

Kuhlmann, S. (1992):
Evaluation von Technologiepolitik. In: Grimmer/Häusler/Kuhlmann/Simonis (Hg.): Politische Techniksteuerung, Opladen

Kuhlmann, S./Berteit, H. (1993):
Die doppelte Herausforderung: Wirtschaftsnahe Forschungs-Infrastruktur Deutschlands im Umbruch, Karlsruhe

Martin, B./Irvine, J. (1986)
An International Comparison of Government Funding of Academic and Academically Related Research. ABRC Science Policy Studies No. 2, Advisory Board for the Research Councils, London

Meyer-Krahmer, F. (1989):
Der Einfluß staatlicher Technologiepolitik auf industrielle Innovationen, Baden-Baden

Meyer-Krahmer, F. (1992):
The German R&D system in transition: Empirical results and prospects of future development. In: Research policy, vol. 21, 5, 423-436

Meyer-Krahmer, F./ Kuntze, U. (1992):
Bestandsaufnahme der Forschungs- und Technologiepolitik. In: Grimmer/ Häusler/ Kuhlmann/ Simonis (Hg.): Politische Techniksteuerung, Opladen

OECD (1992):
Frascati Manual 1992 (Draft 9/1992): The Measurement of Scientific and Technical Activities. Proposed Standard Practice for Surveys of Research and Experimental Development, Paris

PREST/SPRU (1992) (Georghiou, L./Stein, J.-A./Janes, J./Senker, J./Pifer, J./Cameron, H./Neveda, M./Yates, J./Boden, M.):
The Impact of European Community Policies for Research and Technological Development Upon Science and Technology in the United Kingdom (UKIMPACT), Manchester/ Sussex

Starbatty, J./ Vetterlein, U. (1990):
Die Technologiepolitik der Europäischen Gemeinschaft, Baden-Baden

SV-Wissenschaftsstatistik (1990):
Forschung und Entwicklung in Unternehmen nach Größenklassen in den Jahren 1964-1987. Materialien zur Wissenschaftsstatistik, Heft 5, Essen

SV-Wissenschaftsstatistik (1993):
Forschung und Entwicklung in der Wirtschaft - Ergebnisse und Schätzungen 1991 - 92, Informationsblatt, Essen

Tsipouri, L./Gonard, Th./Kuhlmann, S./Morandini, C. (1992):
Analysis of the value added due to multinational university-industry partnerships in EC research projects, Athens et al. (MONITOR-SPEAR study no. 5.7)

Wissenschaftsrat (1988):
Empfehlungen des Wissenschaftsrates zu den Perspektiven der Hochschulen in den 90er Jahren, Köln

Wissenschaftsrat (1993):
Drittmittel der Hochschulen 1970 bis 1993, Köln

Annex : Tables and Figures

Table A 3.4.1: Structure of the German Participants according to EC Programmes and Type of Organisation

EC Programme Count EC Programme % Type of Participant %	Type of Participant					Total
	BIGs	SMEs	HEIs	RDIs	unknown	
AIM	8 23.5% 1.2%	2 5.9% 1.2%	12 35.3% 1.8%	11 32.4% 2.0%	1 2.9% 2.1%	34 100.0% 1.6%
BIO/BRIDGE	3 8.6% .5%		16 45.7% 2.4%	16 45.7% 2.9%		35 100.0% 1.7%
BRITE/EURAM	159 49.8% 24.1%	44 13.8% 26.5%	53 16.6% 7.8%	63 19.7% 11.3%		319 100.0% 15.1%
CODES/STANDARDISATION	15 100.0% 2.3%					15 100.0% .7%
DECOMMISSIONING	11 36.7% 1.7%	7 23.3% 4.2%	6 20.0% .9%	6 20.0% 1.1%		30 100.0% 1.4%
DELTA	7 43.8% 1.1%	2 12.5% 1.2%	2 12.5% .3%	5 31.3% .9%		16 100.0% .8%
DRIVE	42 42.4% 6.4%	7 7.1% 4.2%	30 30.3% 4.4%	12 12.1% 2.2%	8 8.1% 17.0%	99 100.0% 4.7%
ECLAIR	3 27.3% .5%	2 18.2% 1.2%	5 45.5% .7%	1 9.1% .2%		11 100.0% .5%
ESPRIT II	257 38.5% 38.9%	67 10.0% 40.4%	182 27.3% 26.8%	126 18.9% 22.7%	35 5.2% 74.5%	667 100.0% 31.6%
EUROTRA II	1 7.7% .2%		3 23.1% .4%	9 69.2% 1.6%		13 100.0% .6%
FLAIR				1 100.0% .2%		1 100.0% .0%
HUMAN GENOME ANALYSIS				1 100.0% .2%		1 100.0% .0%
JOULE	36 23.5% 5.5%	12 7.8% 7.2%	59 38.6% 8.7%	46 30.1% 8.3%		153 100.0% 7.3%
LARGE SCALE FACILITIES				1 100.0% .2%		1 100.0% .0%

EC Programme Count EC Programme % Type of Participant %	Type of Participant					Total
	BIGs	SMEs	HEIs	RDIs	unknown	
MAST		2 6.9% 1.2%	17 58.6% 2.5%	10 34.5% 1.8%		29 100.0% 1.4%
MEDICINE/HEALTH			8 50.0% 1.2%	8 50.0% 1.4%		16 100.0% .8%
MONITOR			3 30.0% .4%	7 70.0% 1.3%		10 100.0% .5%
RACE I	95 67.4% 14.4%	12 8.5% 7.2%	6 4.3% .9%	25 17.7% 4.5%	3 2.1% 6.4%	141 100.0% 6.7%
RADIATION PROTECTION			18 30.0% 2.7%	42 70.0% 7.6%		60 100.0% 2.8%
RADIOACTIVE WASTE	7 29.2% 1.1%	1 4.2% .6%	2 8.3% .3%	14 58.3% 2.5%		24 100.0% 1.1%
RAW MATERIALS	7 15.2% 1.1%	5 10.9% 3.0%	22 47.8% 3.2%	12 26.1% 2.2%		46 100.0% 2.2%
S.T.D.			24 57.1% 3.5%	18 42.9% 3.2%		42 100.0% 2.0%
SCIENCE			155 68.6% 22.8%	71 31.4% 12.8%		226 100.0% 10.7%
SPES			6 85.7% .9%	1 14.3% .2%		7 100.0% .3%
STEP/EPOCH			47 53.4% 6.9%	41 46.6% 7.4%		88 100.0% 4.2%
TELEMAN	5 38.5% .8%	2 15.4% 1.2%	1 7.7% .1%	5 38.5% .9%		13 100.0% .6%
VALUE	4 36.4% .6%	1 9.1% .6%	2 18.2% .3%	4 36.4% .7%		11 100.0% .5%
Total	660 31.3% 100.0%	166 7.9% 100.0%	679 32.2% 100.0%	556 26.4% 100.0%	47 2.2% 100.0%	2108 100.0% 100.0%

BIGs:	Large firms with more than 500 employees
SMEs:	Small and medium-sized firms with less than 500 employees
HEIs:	Higher education institutes
RDIs:	non-university R&D institutes
Unknown:	due to lack of or insufficient information could not be placed

Table A 3.5.1: Regional Distribution of Participants in the Second Framework Programme

Regions (Regierungsbezirk)	NUTS 2	Participation absolute	Participation in %
Baden-Württemberg			
Freiburg	R183	42	2.0
Karlsruhe	R182	163	7.7
Stuttgart	R181	183	8.7
Tübingen	R184	91	4.3
Bayern			
Mittelfranken	R195	67	3.2
Niederbayern	R192	6	0.3
Oberbayern	R191	332	15.7
Oberfranken	R194	8	0.4
Oberpfalz	R193	7	0.3
Schwaben	R197	16	0.8
Unterfranken	R196	21	1.0
Berlin (Land)	R1B	111	5.3
Bremen (Land)	R14	52	2.5
Hamburg (Land)	R12	101	4.8
Hessen			
Darmstadt	R16A	127	6.0
Gießen	R16B	27	1.3
Kassel	R16C	4	0.2
Niedersachsen			
Braunschweig	R13A	94	4.5
Hannover	R13B	66	3.1
Lüneburg	R13C	4	0.2
Weser-Ems	R13D	17	0.8
Nordrhein-Westfalen			
Arnsberg	R155	64	3.0
Detmold	R154	36	1.7
Düsseldorf	R151	88	4.2
Köln	R152	250	11.9
Münster	R153	20	0.9
Rheinland-Pfalz			
Koblenz	R171	2	0.1
Rheinhessen-Pfalz	R173	37	1.8
Trier	R172	1	0.0
Saarland (Land)	R1A	37	1.8
Sachsen			
Dresden	R1E2	2	0.1
Schleswig-Holstein (Land)	R11	32	1.5
Total		2108	100.0%

Table A 4.1.1: Regional Distribution of Enterprises investigated

Base of Organisation (Region)	Participants		Non-participants	
	absolute	%	absolute	%
Schleswig-Holstein	3	1.4	1	0.4
Hamburg (Land)	10	4.8	13	4.7
Braunschweig	8	3.8	5	1.8
Hannover	3	1.4		
Lüneburg	2	1.0	1	0.4
Weser-Ems	1	0.5	2	0.7
Bremen (Land)	9	4.3	4	1.4
Düsseldorf	14	6.7	26	9.4
Köln	19	9.1	19	6.9
Münster	1	0.5	1	0.4
Arnsberg	1	0.5	4	1.4
Detmold	3	1.4	7	2.5
Darmstadt	23	11.0	3	1.1
Gießen	4	1.9	2	0.7
Kassel	1	0.5	4	1.4
Koblenz	1	0.5		
Rheinhessen-Pfalz	5	2.4		
Stuttgart	18	8.6	26	9.4
Karlsruhe	13	6.2	24	8.7
Freiburg	8	3.8	12	4.3
Tübingen	10	4.8	9	3.2
Oberbayern	34	16.3	32	11.6
Niederbayern			3	1.1
Oberpfalz			1	0.4
Mittelfranken	3	1.4	4	1.4
Unterfranken	2	1.0	10	3.6
Schwaben	5	2.4	13	4.7
Saarland	2	1.0		
Berlin (Land)	6	2.9		
Brandenburg			1	0.4
Mecklenburg-Vorpommern			3	1.1
Chemnitz			4	1.4
Leipzig			14	5.1
Halle			4	1.4
Magdeburg			11	4.0
Thüringen			14	5.1
Total	209	100.0	277	100.0

Table A 5.1.1: Size of the HEIs investigated				
Academic Personnel (full-time employees)	Participants		Non-participants	
	absolute	%	absolute	%
Up to 10	68	35.6	90	48.4
11 - 49	104	54.5	85	45.7
50 - 149	16	8.4	7	3.8
150 and more	3	1.6	4	2.2
Total	191	100.0	186	100.0

Table A 5.1.2: Regional Distribution of HEIs investigated				
Base of Organisation (Region)	Participant		Non-participant	
	absolute	%	absolute	%
Schleswig-Holstein	4	2.0	1	0.5
Hamburg (Land)	10	5.0	5	2.6
Braunschweig	19	9.5	7	3.7
Hannover	11	5. 5	11	5.8
Weser-Ems	2	1.0	2	1.1
Bremen (Land)	8	4.0		
Düsseldorf	13	6.5	11	5.8
Köln	23	11.4	11	5.8
Münster	4	2.0	3	1.6
Arnsberg	14	7.0	13	6.9
Detmold	4	2.0	2	1.1
Darmstadt	9	4.5	8	4.2
Gießen	7	3.5	1	0.5
Kassel			2	1.1
Trier			1	0.5
Rheinhessen-Pfalz	1	0.5	7	3.7
Stuttgart	11	5.5	10	5.3
Karlsruhe	5	2.5	14	7.4
Freiburg	5	2.5	2	1.1
Tübingen	10	5.0	7	3.7
Oberbayern	13	6.5	9	4.8
Oberpfalz	2	1.0	1	0.5
Oberfranken	1	0.5		
Mittelfranken	6	3.0	3	1.6
Unterfranken	4	2.0		
Saarland			2	1.1
Berlin (Land)	9	4.5	5	2.6
Brandenburg			2	1.1
Mecklenburg-Vorpommern			10	5.3
Chemnitz	3	1.5	7	3.7
Dresden	1	0.5	15	7.9
Leipzig	1	0.5	3	1.6
Halle			7	3.7
Magdeburg			2	1.1
Thüringen	1	0.5	5	2.6
Total	201	100	189	100.0

Figure A 5.5.3.1: Has this Project had any further social or environmentally relevant Impacts?

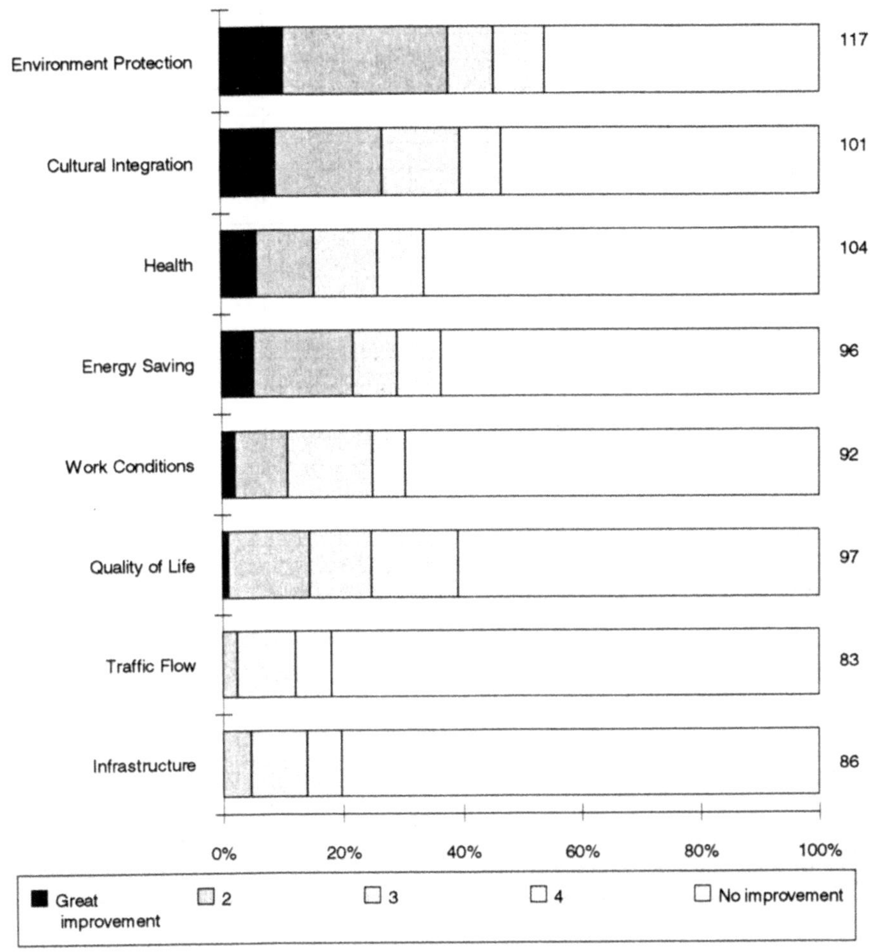

Table A 6.1.3: Regional Distribution of RDIs investigated

Base of Organisation (Region)	Participants		Non-participants	
	absolute	%	absolute	%
Schleswig-Holstein	2	1.1	1	1.7
Hamburg (Land)	8	4.5	1	1.7
Braunschweig	21	11.9	1	1.7
Hannover	5	2.8	1	1.7
Weser-Ems	1	0.6		
Bremen (Land)	2	1.1	1	1.7
Düsseldorf	10	5.7	8	13.3
Köln	17	9.7	3	5.0
Münster	1	0.6	1	1.7
Arnsberg			1	1.7
Detmold	1	0.6		
Darmstadt	12	6.8	2	3.3
Kassel	2	1.1	1	1.7
Koblenz	1	0.6		
Rheinhessen-Pfalz	2	1.1	1	1.7
Stuttgart	11	6.3	2	3.3
Karlsruhe	19	10.8	1	1.7
Freiburg	1	0.6	3	5.0
Tübingen	4	2.3	3	5.0
Oberbayern	24	13.6	2	3.3
Oberpfalz	1	0.6		
Mittelfranken	2	1.1	2	3.3
Unterfranken	1	0.6		
Saarland	7	4.0		
Berlin (Land)	18	10.2	4	6.7
Brandenburg	1	0.6	2	3.3
Mecklenburg-Vorpommern	1	0.6		
Chemnitz	1	0.6	1	1.7
Dresden			6	10.0
Leipzig			4	6.7
Halle			2	3.3
Magdeburg			2	3.3
Thüringen			4	6.7
Total	176	100.0	60	100.0

TECHNOLOGY, INNOVATION, and POLICY

Series of the Fraunhofer Institute
for Systems and Innovation Research (ISI)

Volume 1:
Kerstin Cuhls, Terutaka Kuwahara
Outlook for Japanese and German Future Technology
1994. XVI, 238 pages.
ISBN 3-7908-0800-8

GPSR Compliance

The European Union's (EU) General Product Safety Regulation (GPSR) is a set of rules that requires consumer products to be safe and our obligations to ensure this.

If you have any concerns about our products, you can contact us on ProductSafety@springernature.com

In case Publisher is established outside the EU, the EU authorized representative is:

Springer Nature Customer Service Center GmbH
Europaplatz 3
69115 Heidelberg, Germany

Batch number: 10725973

Printed by Printforce, the Netherlands